SOLUTIONS MANUAL FOR
Numerical Techniques in
Electromagnetics
Second Edition

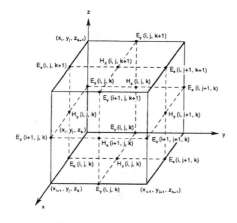

Matthew N. O. Sadiku

CRC Press
Boca Raton   London   New York   Washington, D.C.

TABLE OF CONTENTS

CHAPTER 1

Prob. 1.1

(a)

$$\nabla \times \nabla \Phi = \begin{bmatrix} \frac{\partial}{\partial x} & \frac{\partial}{\partial y} & \frac{\partial}{\partial z} \\ \frac{\partial \Phi}{\partial x} & \frac{\partial \Phi}{\partial y} & \frac{\partial \Phi}{\partial z} \end{bmatrix}$$

$$= \left(\frac{\partial^2 \Phi}{\partial y \partial z} - \frac{\partial^2 \Phi}{\partial z \partial y} \right) \mathbf{a}_x + \left(\frac{\partial^2 \Phi}{\partial x \partial z} - \frac{\partial^2 \Phi}{\partial z \partial x} \right) \mathbf{a}_y + \left(\frac{\partial^2 \Phi}{\partial x \partial y} - \frac{\partial^2 \Phi}{\partial y \partial x} \right) \mathbf{a}_z = 0$$

(b)

$$\nabla \cdot \nabla \times \mathbf{F} = \left(\frac{\partial}{\partial x}, \frac{\partial}{\partial y}, \frac{\partial}{\partial z} \right) \cdot \begin{bmatrix} \frac{\partial}{\partial x} & \frac{\partial}{\partial y} & \frac{\partial}{\partial z} \\ F_x & F_y & F_z \end{bmatrix}$$

$$= \frac{\partial}{\partial x} \left(\frac{\partial F_z}{\partial y} - \frac{\partial F_y}{\partial z} \right) - \frac{\partial}{\partial y} \left(\frac{\partial F_z}{\partial x} - \frac{\partial F_x}{\partial z} \right) + \frac{\partial}{\partial z} \left(\frac{\partial F_y}{\partial x} - \frac{\partial F_x}{\partial y} \right)$$

$$= \frac{\partial^2 F_z}{\partial x \partial y} - \frac{\partial^2 F_y}{\partial x \partial z} - \frac{\partial^2 F_z}{\partial y \partial x} + \frac{\partial^2 F_x}{\partial y \partial z} + \frac{\partial^2 F_y}{\partial z \partial x} - \frac{\partial^2 F_x}{\partial z \partial y} = 0$$

(c)

$$\nabla(\nabla \cdot \mathbf{F}) - \nabla^2 \mathbf{F} = \left[\frac{\partial}{\partial x} \left(\frac{\partial F_x}{\partial x} + \frac{\partial F_y}{\partial y} + \frac{\partial F_z}{\partial z} \right) - \frac{\partial^2 F_x}{\partial x^2} - \frac{\partial^2 F_x}{\partial y^2} - \frac{\partial^2 F_x}{\partial z^2} \right] \mathbf{a}_x$$

$$+ \left[\frac{\partial}{\partial y} \left(\frac{\partial F_x}{\partial x} + \frac{\partial F_y}{\partial y} + \frac{\partial F_z}{\partial z} \right) - \frac{\partial^2 F_y}{\partial x^2} - \frac{\partial^2 F_y}{\partial y^2} - \frac{\partial^2 F_y}{\partial z^2} \right] \mathbf{a}_y$$

$$+ \left[\frac{\partial}{\partial z} \left(\frac{\partial F_x}{\partial x} + \frac{\partial F_y}{\partial y} + \frac{\partial F_z}{\partial z} \right) - \frac{\partial^2 F_z}{\partial x^2} - \frac{\partial^2 F_z}{\partial y^2} - \frac{\partial^2 F_z}{\partial z^2} \right] \mathbf{a}_z$$

$$= \left[\frac{\partial}{\partial y} \left(\frac{\partial F_y}{\partial x} - \frac{\partial F_x}{\partial y} \right) - \frac{\partial}{\partial z} \left(\frac{\partial F_x}{\partial z} - \frac{\partial F_z}{\partial x} \right) \right] \mathbf{a}_x$$

$$+ \left[\frac{\partial}{\partial z} \left(\frac{\partial F_z}{\partial y} - \frac{\partial F_y}{\partial z} \right) - \frac{\partial}{\partial x} \left(\frac{\partial F_y}{\partial x} - \frac{\partial F_x}{\partial y} \right) \right] \mathbf{a}_y$$

$$+ \left[\frac{\partial}{\partial x} \left(\frac{\partial F_x}{\partial z} - \frac{\partial F_z}{\partial x} \right) - \frac{\partial}{\partial y} \left(\frac{\partial F_z}{\partial y} - \frac{\partial F_y}{\partial z} \right) \right] \mathbf{a}_z$$

$$= \nabla \times \nabla \times \mathbf{F}$$

Prob. 1.2 Let $\mathbf{A} = U \nabla V$ and apply Stokes' theorem

$$\oint_L U \nabla V \cdot d\mathbf{l} = \int_S \nabla \times (U \nabla V) \cdot d\mathbf{S}$$

$$= \int_S (\nabla U \times \nabla V) \cdot d\mathbf{S} + \int_S U (\nabla \times \nabla V) \cdot d\mathbf{S}$$

Since $\nabla \times \nabla V = 0$,

$$\oint_L U\nabla V \cdot d\mathbf{l} = \int_S (\nabla U \times \nabla V) \cdot d\mathbf{S}$$

Similarly, we can show that

$$\oint_L V\nabla U \cdot d\mathbf{l} = \int_S (\nabla V \times \nabla U) \cdot d\mathbf{S} = -\int_S (\nabla U \times \nabla V) \cdot d\mathbf{S}$$

Thus,

$$\oint_L U\nabla V \cdot d\mathbf{l} = -\oint_L V\nabla U \cdot d\mathbf{l}$$

as required.

Prob. 1.3 If $\mathbf{J} = 0 = \rho_v$, then Maxwell's equation become

$$\nabla \cdot \mathbf{B} = 0 \tag{1}$$

$$\nabla \cdot \mathbf{D} = 0 \tag{2}$$

$$\nabla \times \mathbf{E} = -\frac{\partial \mathbf{B}}{\partial t} \tag{3}$$

$$\nabla \times \mathbf{H} = \frac{\partial \mathbf{D}}{\partial t} \tag{4}$$

Since $\nabla \cdot \nabla \times \mathbf{A} = 0$ for any vector field $\mathbf{A}$,

$$\nabla \cdot \nabla \times \mathbf{E} = -\frac{\partial \nabla \cdot \mathbf{B}}{\partial t} = 0$$

$$\nabla \cdot \nabla \times \mathbf{H} = \frac{\partial \nabla \cdot \mathbf{D}}{\partial t} = 0$$

showing that (1) and (2) are incorporated in (3) and (4). Thus Maxwell's equations can be reduced to curl equations (3) and (4).

Prob. 1.4 If $\rho_v \neq 0 \neq \mathbf{J}$,

$$\nabla \cdot \epsilon \mathbf{E} = \rho_v$$

$$\nabla \cdot \mu \mathbf{H} = 0$$

$$\nabla \times \mathbf{E} = -\mu\frac{\partial \mathbf{H}}{\partial t}$$

$$\nabla \times \mathbf{H} = \mathbf{J} + \epsilon\frac{\partial \mathbf{E}}{\partial t}$$

$$\nabla \times \nabla \times \mathbf{E} = -\mu\frac{\partial}{\partial t}\nabla \times \mathbf{H} = -\mu\frac{\partial \mathbf{J}}{\partial t} - \mu\epsilon\frac{\partial^2 \mathbf{E}}{\partial t^2}$$

$$\nabla(\nabla \cdot E) - \nabla^2 \mathbf{E} = -\mu\frac{\partial \mathbf{J}}{\partial t} - \frac{1}{c^2}\frac{\partial^2 \mathbf{E}}{\partial t^2}$$

or

$$\nabla^2 \mathbf{E} - \frac{1}{c^2} \frac{\partial^2 \mathbf{E}}{\partial t^2} = \nabla(\rho_v/\epsilon) + \mu \frac{\partial \mathbf{J}}{\partial t}$$

Similarly,

$$\nabla \times \nabla \times \mathbf{H} = \nabla \times \mathbf{J} + \epsilon \frac{\partial}{\partial t} \nabla \times \mathbf{E}$$

$$\nabla(\nabla \cdot H) - \nabla^2 \mathbf{H} = \nabla \times J - \mu\epsilon \frac{\partial^2 \mathbf{H}}{\partial t^2}$$

or

$$\nabla^2 \mathbf{H} - \frac{1}{c^2} \frac{\partial^2 \mathbf{H}}{\partial t^2} = -\nabla \times \mathbf{J}$$

It is assumed that the medium is free space so that the medium is homogeneous and $u = \dfrac{1}{\sqrt{\mu\epsilon}} = c$.

Prob. 1.5 $\nabla \cdot \mathbf{E} = 0, \ \nabla \cdot \mathbf{H} = 0$

$$\nabla \times \mathbf{E} = \begin{bmatrix} \frac{\partial}{\partial x} & \frac{\partial}{\partial y} & \frac{\partial}{\partial z} \\ \\ E_x & E_y & 0 \end{bmatrix} = -\frac{\partial E_y}{\partial z} \mathbf{a}_x + \frac{\partial E_x}{\partial z} \mathbf{a}_y$$

$$= -10k \sin(\omega t + kz)\mathbf{a}_x - 20k \cos(\omega t - kz)\mathbf{a}_y$$

$$\mathbf{H} = -\frac{1}{\mu_o} \int \nabla \times \mathbf{E} \ \partial t$$

$$= \frac{k}{\omega\mu_o} \left[-10\cos(\omega t - kz)\mathbf{a}_x + 20\sin(\omega t - kz)\mathbf{a}_y \right]$$

which is the given $\mathbf{H}$. Since all of Maxwell's equations are satisfied by the fields, they are genuine EM fields.

Prob. 1.6

$$\nabla \times \mathbf{E} = -\mu \frac{\partial \mathbf{H}}{\partial t} \qquad \rightarrow \qquad \frac{\partial \mathbf{H}}{\partial t} = -\frac{1}{\mu_o \epsilon_o} \nabla \times \epsilon_o \mathbf{E}$$

$$\frac{\partial \mathbf{H}}{\partial t} = -\frac{1}{\mu_o \epsilon_o} \nabla \times \mathbf{D} = -\frac{1}{\mu_o \epsilon_o} \begin{bmatrix} \frac{\partial}{\partial x} & \frac{\partial}{\partial y} & \frac{\partial}{\partial z} \\ \\ D_x(z,t) & 0 & 0 \end{bmatrix}$$

$$= -\frac{1}{\mu_o \epsilon_o} \frac{\partial D_x}{\partial z} \mathbf{a}_y$$

$$= \frac{\beta}{\mu_o \epsilon_o} D_o \sin(\omega t + \beta z)\mathbf{a}_y$$

$$\mathbf{H} = -\frac{D_o}{\beta} \cos(\omega t - \beta z)\mathbf{a}_y$$

Prob. 1.7

$$\nabla \times \mathbf{H}_s = j\omega \mathbf{E}_s = \frac{1}{\rho}\frac{d}{d\rho}(\rho H_\phi)\mathbf{a}_z$$

$$= \frac{H_o}{\rho}\left(\frac{1}{\rho}\rho^{1/2}e^{-j\beta\rho} - j\beta\rho^{1/2}e^{-j\beta\rho}\right)\mathbf{a}_z$$

$$= \frac{H_o}{\sqrt{\rho}}(\frac{1}{\rho^2} - j\beta)e^{-j\beta\rho}\mathbf{a}_z$$

$$\mathbf{E}_s = \frac{\nabla \times \mathbf{H}_s}{j\omega\epsilon} = \frac{1}{j\omega\epsilon}\frac{H_o}{\sqrt{\rho}}(\frac{1}{\rho^2} - j\beta)e^{-j\beta\rho}\mathbf{a}_z$$

Prob. 1.8

$$\nabla \times \mathbf{H} = \frac{\partial \mathbf{D}}{\partial t} = \epsilon\frac{\partial \mathbf{E}}{\partial t} \quad \rightarrow \quad \nabla \times \mathbf{H}_s = j\omega\epsilon\mathbf{E}_s$$

$$\nabla \times \mathbf{H}_s = \frac{1}{r\sin\theta}\frac{IL}{4\pi r}(2\sin\theta\cos\theta)\left(\frac{1}{r} + j\beta\right)e^{-j\beta r}\mathbf{a}_r$$
$$- \frac{IL\sin\theta}{4\pi r}\left[-j\beta(\frac{1}{r} + j\beta)e^{-j\beta r} - \frac{1}{r^2}e^{-j\beta r}\right]\mathbf{a}_\phi$$

$$\mathbf{E}_s = \frac{IL}{4\pi r j\omega\epsilon}e^{-j\beta r}\left[2\cos\theta(\frac{1}{r^2} + \frac{j\beta}{r^2})\mathbf{a}_r - -\sin\theta(\beta^2 - \frac{j\beta}{r} - \frac{1}{r^2})\mathbf{a}_\phi\right]$$

Prob. 1.9

$$E_s = \frac{20(e^{jk_x x} - e^{-jk_x x})}{2j}\frac{(e^{jk_y y} + e^{-jk_y y})}{2j}$$
$$= j5\left[e^{j(k_x x + k_y y)} + e^{j(k_x x - k_y y)} - e^{j(k_x x - k_y y)} - e^{-j(k_x x + k_y y)}\right]$$

which consists of four plane waves. $\nabla \times \mathbf{E}_s = -j\omega\mu_o\mathbf{H}_s$ Or

$$H_s = \frac{j}{\omega\mu_o}\nabla \times \mathbf{E}_s = \frac{j}{\omega\mu_o}\begin{bmatrix}\frac{\partial}{\partial x} & \frac{\partial}{\partial y} & \frac{\partial}{\partial z}\\ 0 & 0 & E_{sz}(x,y)\end{bmatrix}$$

$$= \frac{j}{\omega\mu_o}\left(\frac{\partial E_{sz}}{\partial y}\mathbf{a}_x - \frac{\partial E_{sz}}{\partial x}\mathbf{a}_y\right)$$

$$= -\frac{20}{\omega\mu_o}\left[k_y\sin(k_x x)\sin(k_y y)\mathbf{a}_x + k_x\cos(k_x x)\cos(k_y y)\mathbf{a}_y\right]$$

Prob. 1.10 (a) $I = Re(I_s e^{j\omega t}) = \sin\pi x\cos\pi y\cos(\omega t - z)$
(b) $V = Re(20e^{-j2x}e^{-j90°}e^{j\omega t} - 10e^{-j4x}e^{j\omega t})$

$$V_s = 20e^{-j2x}e^{-j90°} - 10e^{-j4x} = -j20e^{-j2x} - 10e^{-j4x}$$

Prob. 1.11

(a) $A = Re(A_s e^{j\omega t})$

$$A = \cos(\omega t - 2z)\mathbf{a}_x - \sin(\omega t - 2z)\mathbf{a}_y$$

(b) $B = Re(B_s e^{j\omega t})$

$$B = -10\sin x \sin \omega t \mathbf{a}_x - 5\cos(\omega - 2z + 45°)\mathbf{a}_z$$

(c) $C = Re(C_s e^{j\omega t})$

$$C = 2\cos 2x \sin(\omega t - 3x) + e^{3x}\cos(\omega t - 4x)$$

Prob. 1.12 Assuming the time factor $e^{j\omega t}$, equation

$$\nabla^2 \mathbf{E} - \mu\epsilon\frac{\partial^2 \mathbf{E}}{\partial t^2} = \mu\frac{\partial \mathbf{J}}{\partial t}$$

becomes $\nabla^2 \mathbf{E}_s + \omega^2\mu\epsilon\mathbf{E}_s = j\omega\mu\sigma\mathbf{E}_s$ or $\nabla^2 \mathbf{E}_s - j\omega\mu(\sigma+j\omega\epsilon)\mathbf{E}_s = 0$ For conducting medium, $\sigma >> \omega\epsilon$ so that
$$\nabla^2 \mathbf{E}_s - j\omega\mu\sigma\mathbf{E}_s = 0$$

Prob. 1.13 Assuming that $\mathbf{J} = 0 = \rho_v$, Maxwell's equations become

$$\nabla \cdot \epsilon\mathbf{E}_s = 0 \tag{1}$$
$$\nabla \cdot \mu\mathbf{H}_s = 0 \tag{2}$$
$$\nabla \times \mathbf{E}_s = -j\omega\mu\mathbf{H}_s \tag{3}$$
$$\nabla \times \mathbf{H}_s = j\omega\epsilon\mathbf{E}_s \tag{4}$$

From (3), $\mathbf{H}_s = -\frac{1}{j\omega\mu}\nabla \times \mathbf{E}_s$,

$$\nabla \times \mathbf{H}_s = -\nabla \times (\frac{1}{j\omega\mu}\nabla \times \mathbf{E}_s) = j\omega\epsilon\mathbf{E}_s$$

or

$$\nabla \times (\frac{1}{j\omega\mu}\nabla \times \mathbf{E}_s) + j\omega\epsilon\mathbf{E}_s = 0$$

From (4), $\mathbf{E}_s = \frac{1}{j\omega\epsilon}\nabla \times \mathbf{H}_s$,

$$\nabla \times \mathbf{E}_s = \nabla \times (\frac{1}{j\omega\epsilon}\nabla \times \mathbf{H}_s) = -j\omega\mu\mathbf{H}_s$$

or

$$\nabla \times (\frac{1}{j\omega\epsilon}\nabla \times \mathbf{H}_s) + j\omega\mu\mathbf{H}_s = 0$$

Prob. 1.14 Let $\mathbf{A} = Re(A_s e^{j\omega t})$, $V = Re(V_s e^{j\omega t})$, etc.

$$\frac{\partial A}{\partial t} \quad \rightarrow \quad j\omega A_s$$

$$\frac{\partial^2 A}{\partial t^2} \quad \rightarrow \quad -\omega^2 A_s$$

Similarly,

$$\frac{\partial^2 V}{\partial t^2} \quad \rightarrow \quad -\omega^2 V_s$$

Substituting these into eqs. (1.42) and (1.43), we obtain

$$\nabla^2 V_s + \omega^2 \mu\epsilon V_s = -\frac{\rho_{vs}}{\epsilon}$$
$$\nabla^2 \mathbf{A}_s + \omega^2 \mu\epsilon \mathbf{A}_s = -\mu\mathbf{J}_s$$

But $k^2 = \omega^2\mu\epsilon$. Thus

$$\nabla^2 V_s + k^2 V_s = -\frac{\rho_{vs}}{\epsilon}$$
$$\nabla^2 \mathbf{A}_s + k^2 \mathbf{A}_s = -\mu\mathbf{J}_s$$

Prob. 1.15

(a) $a = 1$, $b = 2$, $c = 5$,
$$b^2 - 4ac = -16$$

Hence it is elliptic.

(b) $a = y^2 + 1$, $b = 0$, $c = x^2 + 1$,
$$b^2 - 4ac = -4(x^2 + 1)(y^2 + 1) < 0$$

Hence it is elliptic.

(c) $a = 1$, $b = -2\cos x$, $c = -(3 + \sin^2 x)$,
$$b^2 - 4ac = 4\cos^2 x + 12 + 4\sin^2 x = 16 > 0$$

Hence it is hyperbolic.

(d) $a = x^2$, $b = -2xy$, $c = y^2$,
$$b^2 - 4ac = 4x^2y^2 - 4x^2y^2 = 0$$

Hence it is parabolic.

Prob. 1.16

(a) $a = \alpha$, $b = 0$, $c = 0$, $b^2 - 4ac = 0$, i.e. it is parabolic.

(b) $a = 1$, $b = 0$, $c = 1$, $b^2 - 4ac = -4$, i.e. it is elliptic.

(c) $a = 1$, $b = 0$, $c = 1$, $b^2 - 4ac = -4$, i.e. it is elliptic.

(d)

$$\left(\frac{\partial^2 \Phi}{\partial x^2} + \frac{\partial^2 \Phi}{\partial y^2} \right)^2 = 0$$

$a = 1$, $b = 0$, $c = 1$, $b^2 - 4ac = -4$, i.e. it is elliptic.

CHAPTER 2

Prob. 2.1 Let $\Phi(x,y) = X(x)Y(y)$.

$$aX''Y + bX'Y' + cXY'' + dX'Y + eXY' + fXY = 0$$

Dividing through by XY,

$$\frac{1}{X}(aX'' + dX') + \frac{1}{Y}(cY'' + eY') + \frac{bX'Y'}{XY} + f = 0$$

The PDE is separable if and only if a and d are functions of x only, c and e are functions of y only, $b = 0$, and f is a sum of a function of x only and a function of y only, i.e. if

$$a(x)\Phi_{xx} + c(y)\Phi_{yy} + d(x)\Phi_x + e(y)\Phi_y + [f_1(x) + f_2(y)]\Phi = 0$$

Prob. 2.2 (a) Consider the problem shown below.

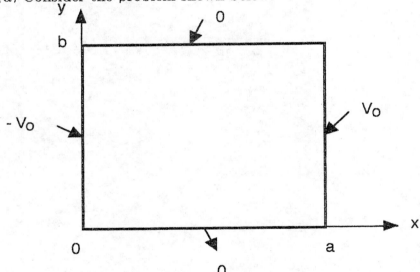

This is similar to Example 2.1 with $V_1 = V_3 = 0$, $V_2 = V_o$, $V_4 = -V_o$. Hence

$$V(x,y) = \frac{4V_o}{\pi} \sum_{n=\text{odd}}^{\infty} \frac{\sinh n\pi x/b \sin n\pi y/b}{n \sinh n\pi a/b} - \frac{4V_o}{\pi} \sum_{n=\text{odd}}^{\infty} \frac{\sinh n\pi(a-x)/b \sin n\pi y/b}{n \sinh n\pi a/b}$$

But $\sinh A - \sinh B = 2\cosh \dfrac{A+B}{2} \sinh \dfrac{A-B}{2}$

$$V(x,y) = \frac{4V_o}{\pi} \sum_{n=\text{odd}}^{\infty} \frac{\sin n\pi y/b}{n \sinh n\pi a/b} 2\cosh \frac{n\pi a}{2b} \sinh \frac{n\pi}{b}(x - a/2)$$

We now transform coordinates:

$$x = X + a/2, \qquad y = Y + b/2$$

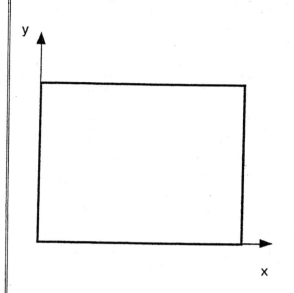
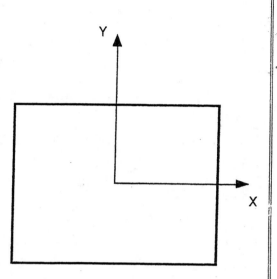

$$V(x,y) = \frac{4V_o}{\pi} \sum_{n=odd}^{\infty} \frac{\sin \frac{n\pi}{b}(Y+b/2)}{2n \sinh \frac{n\pi a}{2b} \cosh \frac{n\pi a}{2b}} 2 \cosh \frac{n\pi a}{2b} \sinh \frac{n\pi X}{b}$$

But

$$\sin(\frac{n\pi Y}{b} + \frac{n\pi}{2}) = \sin \frac{n\pi Y}{b} \cos n\pi/2 + \sin n\pi/2 \cos \frac{n\pi Y}{b}$$
$$= (-1)^n \cos n\pi Y/b, \qquad n = odd$$

$$V(x,y) = \frac{4V_o}{\pi} \sum_{n=odd}^{\infty} \frac{(-1)^n \sinh n\pi x/b \cos n\pi y/b}{n \sinh \frac{n\pi a}{2b}}$$

(b) Let $V(x,y) = X(x)Y(y)$

$$Y_n(x) = \sin \frac{n\pi y}{b}$$
$$X_n(x) = A_1 e^{-n\pi x/b} + A_2 e^{n\pi x/b}$$

$A_2 = 0$ since $X(x) = 0$ as $x \to \infty$.

$$V(x,y) = \sum a_n \sin n\pi y/b \, e^{-n\pi x/b}$$

$$V(0, y) = V_o = \sum a_n \sin n\pi y/b$$

$$a_n = \frac{2}{b} \int_0^b V_o \sin n\pi y/b\, dx = \begin{cases} 0, & n = \text{even} \\ \frac{4V_o}{n\pi}, & n = \text{odd} \end{cases}$$

$$V(x, y) = \frac{4V_o}{\pi} \sum_{n=\text{odd}}^{\infty} \frac{1}{n} \sin n\pi y/b\, e^{-n\pi x/b}$$

Prob. 2.3 (a) This is similar to Example 2.1 with $V_2 = V_3 = V_4 = 0$, $V_1 = V_o a/x$. Hence

$$V = \sum_{n=\text{odd}}^{\infty} A_n \sin(n\pi x/a) \sinh[n\pi(b-y)/a]$$

When $y = 0$, we obtain

$$V_o x/a = \sum_{n=\text{odd}}^{\infty} A_n \sin(n\pi x/a) \sinh(n\pi b/a)$$

Multiplying both sides by $\sin(n\pi x/a)$ and integrating over $0 < x < a$ leads to

$$A_n = \frac{2}{a \sinh(n\pi b/a)} \int_0^a \frac{V_o x}{a} \sin(n\pi x/a)dx$$

$$= \frac{2V_o}{a^2 \sinh(n\pi b/a)} \left[\frac{a^2}{n^2\pi^2} \sin(n\pi x/a) - \frac{ax}{n\pi} \cos(n\pi x/a) \right]_0^a$$

$$= \frac{2V_o(-1)^{n+1}}{n\pi \sinh(n\pi b/a)}$$

(b) Let $V(x, y) = X(x)Y(y)$.

$$X(x) = A_1 \cos \beta x + A_2 \sin \beta x$$

Since $X(-a/2) = 0 = X(a/2)$, $A_2 = 0$ and

$$0 = A_1 \cos \beta a/2 \qquad \rightarrow \qquad n\pi - \pi/2 = \beta a/2$$

or $\beta = \pi(2n-1)/a$. Thus

$$X(x) = A_1 \cos \beta x, \qquad Y(y) = B_1 \cosh \beta y + B_2 \sinh \beta y$$

$$V(x, y) = \sum \cos \beta x \left(C_n \cosh \beta y + D_n \sinh \beta y \right)$$

Substituting $V(x, b/2) = V_o \cos \pi x/a$ and $V(x, -b/2) = V_o \cos \pi x/a$ gives $n = 1$, $D_n = 0$ and

$$C_n = \frac{V_o}{\cosh(\pi b/a)}$$

Thus

$$V = \frac{V_o \cos(\pi x/a) \cosh(\pi y/a)}{\cosh(\pi b/a)}$$

Prob. 2.4 (a) Let $U(x,y) = X(x)Y(y)$

$$X = A_1 \cos \beta x + B_1 \sin \beta x$$

$$X'(0) = 0 \quad \rightarrow \quad B_1 = 0$$
$$X'(\pi) = 0 \quad \rightarrow \quad 0 = -A_1 \beta \sin \beta \pi$$

i.e. $\beta = n, \quad n = 1, 2, 3, \cdots$

$$X = A_1 \cos nx, \qquad Y = A_2 \cosh \beta y + B_2 \sinh \beta y$$

$$Y(0) = 0 \quad \rightarrow A_2 = 0$$

Hence

$$U(x,y) = C_o + \sum_{n=1}^{\infty} C_n \cos nx \sinh ny$$

$$= \sum_{n=0}^{\infty} C_n \cos nx \sinh ny$$

$$U(x,\pi) = x = \sum_{n=0}^{\infty} C_n \cos nx \sinh n\pi$$

Multiplying both sides by $\cos mx$ and integrating over $0 < x < \pi$ yields $C_o = \pi/2$ and

$$C_n = \begin{cases} 0, & n = \text{even} \\ \dfrac{4}{n\pi^2 \sinh n\pi}, & n = \text{odd} \end{cases}$$

$$U(x,y) = \frac{\pi}{2} - \frac{4}{\pi} \sum_{n=\text{odd}}^{\infty} \frac{\cos nx \sinh ny}{n^2 \sinh n\pi}$$

(b) First, transform the equation by letting

$$U(x,t) = x + u(x,t)$$

where

$$u_t = k u_{xx}$$

subject to

$$u(0,t) = 0 = u(1,t), \ u(x,0) = -x$$

If $u(x,t) = X(x)T(t)$,

$$X = A_1 \cos \beta x + A_2 \sin \beta x$$

$$X(0) = 0 \quad \rightarrow \quad A_1 = 0$$
$$X(1) = 0 \quad \rightarrow \quad 0 = \sin \beta$$

i.e. $\beta = n\pi$.

$$X(x) = A_2 \sin n\pi x, \qquad T = Be^{-kn^2\pi^2 t}$$

$$u(x,t) = \sum_{n=1}^{\infty} C_n \sin n\pi x e^{-kn^2\pi^2 t}$$

$$u(x,0) = -x = \sum_{n=1}^{\infty} C_n \sin n\pi x$$

$$C_n = -2 \int_0^1 x \sin n\pi x \, dx = \frac{2(-1)^n}{n\pi}$$

$$U(x,t) = x + u(x,t)$$

$$= x + \frac{2}{\pi} \sum_{n=1}^{\infty} \frac{(-1)^n}{n} \sin n\pi x e^{-kn^2\pi^2 t}$$

(c) Let $u(x,t) = X(x)T(t)$. From the boundary conditions,

$$X = A \cos \beta x + B \sin \beta x$$

$$X(0) = 0 \quad \rightarrow \quad A = 0$$
$$X(1) = 0 \quad \rightarrow \quad 0 = \sin \beta$$

or $\beta = n\pi$.

$$X = A \sin n\pi x, \qquad T = A_3 \sin n\pi a t + A_4 \cos n\pi a t$$

$$T(0) = 0 \quad \rightarrow \quad A_4 = 0$$

$$u(x,t) = \sum_{n=1}^{\infty} C_n \sin n\pi x \sin n\pi a t$$

$$u_t(x,0) = x = \sum_{n=1}^{\infty} n\pi a C_n \sin n\pi x$$

Multiplying both sides by $\sin m\pi x$ and integrating over $0 < x < 1$ yields

$$C_n = -\frac{2\cos n\pi a}{(n\pi a)^2}$$

$$u(x,t) = -\frac{2}{\pi^2 a^2}\sum_{n=1}^{\infty}\frac{\cos n\pi a}{n^2}\sin n\pi x \sin n\pi a t$$

Prob. 2.5 (a) Let $\Phi(\rho,\phi) = R(\rho)F(\phi)$. From the text, eq. (2.48) onward,

$$F(\phi) = C_1\cos(\lambda\phi) + C_2\sin(\lambda\phi)$$

$C_1 = 0$ since $\Phi(\rho,0) = \Phi(\rho,\pi)$. Also, $n = 1, 2, \cdots$.

$$F_n(\phi) = C_n\sin n\phi$$

$$R_n(\phi) = a_n\rho^n + b_n\rho^{-n}$$

As $\rho \to 0$, $\Phi(\rho,\phi)$ must be finite, i.e. $a_n = 0$.

$$\Phi(\rho,\phi) = \sum_{n=1}^{\infty}A_n\frac{\sin n\phi}{\rho^n}$$

$$\Phi(1,\phi) = \sin\phi = \sum_{n=1}^{\infty}A_n\sin n\phi$$

which implies that $A_1 = 1$, $A_n = 0$ if $n \neq 1$. Thus

$$\Phi(\rho,\phi) = \frac{\sin\phi}{\rho}$$

(b) From the boundary conditions, Φ does not depend on ϕ, i.e. $m = 0$ in $F'' + m^2 F = 0$. From eq. (2.68) in the text,

$$Z(z) = C_1\sin\mu z + C_2\cos\mu z$$

If $Z(0) = 0 = Z(L)$, then $C_2 = 0$ and $\mu L = n\pi$ or $\mu = n\pi/L$. Also,

$$\rho^2 R'' + \rho R' - \mu^2\rho^2 R = 0$$

or

$$\rho^2 R'' + \rho R' + (j^2\mu^2\rho^2 - 0)R = 0$$

The solution is

$$R(\rho) = A_n I_0(\rho\mu) + B_n K_0(\rho\mu)$$

$B_n = 0$ since K_o is infinite at $\rho = 0$. Thus

$$\Phi(\rho, z) = \sum_{n=1}^{\infty} A_n I_o(n\pi\rho/L) \sin(n\pi z/L)$$

To obtain A_n, we apply the boundary condition $\Phi(a, z) = 1$, multiply both sides by $\sin m\pi z/L$, and integrate over $0 < z < L$. We get

$$A_n = \begin{cases} 0, & n = \text{even} \\ \frac{4}{n\pi I_0(n\pi a/L)}, & n = \text{odd} \end{cases}$$

$$\Phi(\rho, \phi) = \frac{4}{\pi} \sum_{n=1,3,5}^{\infty} \frac{I_0(n\pi\rho/L)}{I_0(n\pi a/L)} \frac{\sin(n\pi z/L}{n}$$

(c) Let $\Phi(\rho, \phi, t) = R(\rho)F(\phi)T(t)$. Separation of variables leads to

$$T' + \mu^2 kT = 0 \qquad \rightarrow \qquad T(t) = e^{-k\mu^2 t}$$

From eq. (2.67) in the text,

$$F_n(\phi) = a_n \cos n\phi + b_n \sin n\phi$$

Since $\Phi(\rho, \phi, 0) = \rho \cos\phi$, $b_n = 0$ and

$$F(\phi) = a_n \cos n\phi$$

Finally,

$$\rho^2 R'' + R' + (\mu^2 \rho^2 - n^2)R = 0$$

with solution

$$R(\rho) = c_1 J_n(\rho\mu) + c_2 Y_n(\rho\mu)$$

$c_2 = 0$ since R must be finite at $\rho = 0$. Hence

$$\Phi(\rho, \phi, t) = \sum_\mu \sum_n A_{n\mu} J_n(\rho\mu) \cos n\phi e^{-k\mu^2 t}$$

But $\Phi(a, \phi, t) = 0$ implies that

$$J_n(\mu a) = 0 = J_n(X_m)$$

where X_m are the roots of J_n and $\mu = X_m/a$. Also,

$$\Phi(\rho, \phi, 0) = \rho^2 \cos 2\phi = \sum_\mu \sum_n A_{n\mu} J_n(\rho\mu) \cos n\phi$$

15

It is evident that $n = 2$ and that

$$\rho^2 = \sum_\mu A_{2\mu} J_2(\rho\mu)$$

which is the Fourier-Bessel expansion of ρ^2. From Table 2.1, property (h),

$$A_{2\mu} = \frac{2}{a^2[J_3(a\mu)]^2} \int_0^a \rho^2 J_2(\rho\mu)d\rho = \frac{2}{X_m J_3(X_m)}$$

Thus

$$\Phi(\rho, \phi, t) = 2 \sum_{m=1}^\infty \frac{J_2(\rho X_m/a)}{X_m J_3(X_m)} \cos 2\phi \exp(-X_m^2 kt/a^2)$$

where $J_2(X_m) = 0$.

Prob. 2.6 Let $\Phi_{xxxx} + a\Phi_{tt} = 0$, $a = 1/4$, and

$$\Phi = X(x)T(t) \tag{1}$$

The general solution for X is

$$X = A \sinh \lambda x + B \cosh \lambda x + C \sin \lambda x + D \cos \lambda x \tag{2}$$

Applying the boundary conditions,

$$X(0) = 0 \quad \rightarrow \quad B + D = 0 \tag{3}$$
$$X''(0) = 0 \quad \rightarrow \quad B - D = 0 \tag{4}$$
$$X(1) = 0 \quad \rightarrow \quad A \sinh \lambda + B \cosh \lambda + C \sin \lambda + D \cos \lambda \tag{5}$$
$$X''(1) = 0 \quad \rightarrow \quad A \sinh \lambda + B \cosh \lambda - C \sin \lambda - D \cos \lambda \tag{6}$$

From (3) to (6), $A = 0 = B = D$, $\lambda = n\pi$, $n = 1, 2, 3, \cdots$ Thus,

$$X_n(x) = c_n \sin n\pi x$$

For T,

$$T_n(t) = E_n \sin a\lambda^2 t + F_n \cos a\lambda^2 t$$
$$T'(0) = 0 \quad \rightarrow \quad E_n$$

Hence,

$$\Phi(x, t) = \sum_{n=1}^\infty g_n \sin n\pi x \cos an^2\pi^2 t$$

Imposing

$$\Phi(x, 0) = x = \sum_{n=1}^\infty g_n \sin n\pi x$$

16

leads to

$$g_n = \frac{2(-1)^{n+1}}{n\pi}$$

$$\Phi(x,t) = \sum_{n=1}^{\infty} \frac{2(-1)^{n+1}}{n\pi} \sin n\pi x \cos n^2\pi^2 t/4$$

Prob. 2.7 If we let $V(\rho, z) = R(\rho)Z(z)$, after separation of variables, we obtain

$$R = AI_o(k\rho) + BK_0(k\rho)$$
$$Z = C\cos kz + D\sin kz$$

$B = 0$ since V is finite at $\rho = 0$. Also, since $V(\rho, 0) = 0 = V(\rho, L)$, $C = 0$, $k = n\pi/L$, $n = 1, 2, \cdots$

$$V(\rho, z) = \sum_{n=1}^{\infty} A_n \sin(n\pi z/L)I_0(n\pi\rho/L)$$

Imposing the last condition,

$$V(a, z) = \sum_{n=1}^{\infty} A_n \sin(n\pi z/L)I_0(n\pi a/L)$$

Multiplying by $\sin(n\pi z/L)$ and integrating over $0 < z < L$ gives

$$A_n I_0(n\pi a/L) = \frac{2}{L}\int_0^L V(a,z)\sin(n\pi z/L)dz$$
$$= \frac{2V_o}{L}\left[\int_0^{L/2}\frac{z}{L}\sin(n\pi z/L)dz + \int_{L/2}^L (1 - z/L)\sin(n\pi z/L)dz\right]$$

Thus,

$$A_n = \frac{2V_o}{n^2\pi^2 I_0(n\pi a/L)}(2\sin n\pi/2 - n\pi\cos n\pi/2) - \frac{2V_o}{n\pi I_0(n\pi a/L)}(\cos n\pi - \cos n\pi/2)$$

and

$$V(\rho, z) = \sum_{n=1}^{\infty} A_n \sin(n\pi z/L)I_0(n\pi\rho/L)$$

Substituting $\rho = 0.8a$ and $z = 0.3L$ leads to

$$V(0.8a, 0.3L) = 0.26396V_o$$

Prob. 2.8 Let $V(\rho, z) = R(\rho)Z(z)$. After separation of variables, we get

$$Z'' + k^2 Z = 0, \qquad \rho^2 R'' + \rho R - (k\rho)^2 R = 0$$

which have solutions

$$R = AI_o(k\rho) + BK_0(k\rho), \ Z = C\cos kz + D\sin kz$$

$B = 0$ since the potential is finite at $\rho = 0$. Also, since $V(\rho, 0) = 0 = V(\rho, L)$, $C = 0$, $k = n\pi/L$, $n = 1, 2, 3, \cdots$.

$$V(\rho, z) = \sum_{n=1}^{\infty} A_n \sin(n\pi z/L) I_0(n\pi\rho/L)$$

$$V(a, z) = V_o = \sum_{n=1}^{\infty} A_n \sin(n\pi z/L) I_0(n\pi a/L)$$

$$A_n I_0(n\pi a/L) = \frac{2}{L} \int_0^L V_o \sin(n\pi z/L) dz = \begin{cases} 0, & n = \text{even} \\ \frac{4V_o}{n\pi}, & n = \text{odd} \end{cases}$$

Thus,

$$V(\rho, z) = \frac{4V_o}{\pi} \sum_{n=\text{odd}}^{\infty} \frac{I_0(n\pi\rho/L)}{n I_0(n\pi a/L)} \sin(n\pi z/L)$$

Prob. 2.9 Let $V(\rho, \phi) = R(\rho)F(\phi)$, $F(\phi) = c_1 \cos(\lambda\phi) + c_2 \sin(\lambda\phi)$.

$$F(0) = 0 \qquad \rightarrow \qquad c_1 = 0$$
$$F(\pi/3) = 0 \qquad \rightarrow \qquad c_2 \sin \lambda\pi/3 = 0$$

Thus $\lambda\pi/3 = n\pi$ or $\lambda = 3n$.

$$F(\phi) = c_2 \sin(3n\phi), \qquad R(\rho) = c_3\rho^{3n} + c_4\rho^{-3n}$$

But $R(a) = 0$ gives $c_4 = c_3 a^{6n}$. Hence

$$R(\rho) = A_3\left[(\rho/a)^{3n} - (\rho/a)^{-3n}\right]$$

Thus,

$$V(\rho, \phi) = \sum_{n=0}^{\infty} A_n\left[(\rho/a)^{3n} - (\rho/a)^{-3n}\right] \sin 3n\phi$$

$$V(b, \phi) = \sum_{n=0}^{\infty} A_n\left[(b/a)^{3n} - (b/a)^{-3n}\right] \sin 3n\phi$$

18

Multiplying both sides by $\sin 3m\phi$ and integrating over $0 < \phi < \pi/3$ gives

$$A_n = \begin{cases} 0, & n = \text{even} \\ \dfrac{4V_0}{n\pi\left[(b/a)^{3n} - (b/a)^{-3n}\right]}, & n = \text{odd} \end{cases}$$

$$V(\rho,\phi) = \frac{4V_0}{\pi} \sum_{n=1,3,5}^{\infty} \frac{1}{n} \frac{\left[(\rho/a)^{3n} - (\rho/a)^{-3n}\right]}{\left[(\rho/a)^{3n} - (\rho/a)^{-3n}\right]} \sin 3n\phi$$

Prob. 2.10 Let $\Phi(\rho,t) = F(\rho)T(t)$. Separation of variables gives

$$\rho F'' + F' + \lambda^2 \rho F, \qquad T'' + \lambda^2 T = 0$$

F is Bessel's function of order zero.

$$F = c_1 J_0(\lambda\rho) + c_2 Y_0(\lambda\rho)$$

Since F is bounded, $c_2 = 0$. Also,

$$F(a) = 0 \qquad \rightarrow \qquad J_0(\lambda a) = 0 = J(\lambda_{0n})$$

or $\lambda = \lambda_{0n}/a$. Thus

$$\Phi = \sum_{n=0}^{\infty} J_0(\lambda_{0n}\rho/a)\left[A_n \cos(\lambda_{0n}t/a) + B_n \sin(\lambda_{0n}t/a)\right]$$

$$\Phi'(\rho,0) = 0 \qquad \rightarrow \qquad B_n = 0$$

Also,

$$\Phi(\rho,0) = 1 - \rho^2/a^2 = \sum_{n=0}^{\infty} A_n J_0(\lambda_{0n}\rho/a)$$

which is Fourier-Bessel expansion over $[1,0]$ for $(1-x^2)$, where $x = \rho/a$.

$$A_n = \frac{2}{J_1^2(\lambda_{0n})} \int_0^1 (1-x^2)x J_0(\lambda_{0n}x)dx$$

Integrating by parts leads to

$$A_n = \frac{4J_2(\lambda_{0n})}{\lambda_{0n}^2 J_1(\lambda_{0n})}$$

and

$$\Phi(\rho,t) = \sum_{n=0}^{\infty} \frac{4J_2(\lambda_{0n})}{\lambda_{0n}^2 J_1(\lambda_{0n})} J_0(\lambda_{0n}\rho/a) \cos(\lambda_{0n}t/a)$$

Prob. 2.11 (a)

$$\frac{\partial G}{\partial x} = \frac{1}{2}(t - 1/t)\exp[\frac{x}{2}(t - 1/t)]$$

$$= \sum t^n \frac{d}{dx}J_n$$

But

$$\frac{1}{2}(t - 1/t)\exp[\quad] = \frac{1}{2}t\exp[\quad] - \frac{1}{2t}\exp[\quad]$$

$$= \frac{1}{2}\sum t^{m+1}J_m - \frac{1}{2}\sum t^{k-1}J_k$$

Substituting $m = n + 1$ in the right term on the right hand side and $n = k - 1$ in the second term, we obtain

$$\sum t^n \frac{d}{dx}J_n = \frac{1}{2}\sum t^n J_{n-1} - \frac{1}{2}\sum t^n J_{n+1}$$

Hence

$$\frac{d}{dx}J_n(x) = \frac{1}{2}[J_{n-1}(x) - J_{n+1}(x)]$$

(b)

$$\frac{\partial G}{\partial t} = \frac{x}{2}(1 + \frac{1}{t^2})\exp[\frac{x}{2}(t - 1/t)]$$

$$= \sum n t^{n-1}J_n$$

But

$$\frac{x}{2}(1 + \frac{1}{t^2})\exp[\frac{x}{2}(t - 1/t)] = \frac{x}{2}\exp[\quad] + \frac{x}{2t^2}\exp[\quad]$$

$$= \frac{x}{2}\sum t^m J_m + \frac{x}{2}\sum t^{k-2}J_k(x)$$

Hence

$$\sum n t^{n-1}J_n = \frac{x}{2}\sum t^m J_m + \frac{x}{2}\sum t^{k-2}J_k(x)$$

We replace $n - 1$ by m on the left hand side and $k - 2$ by m in the second term on the right hand side.

$$\sum (m + 1)t^m J_{m+1}(x) = \frac{x}{2}\sum t^m J_m + \frac{x}{2}\sum t^m J_{m+2}(x)$$

Thus

$$(m + 1)J_{m+1} = \frac{x}{2}[J_m + J_{m+1}]$$

or

$$J_{n+1}(x) = \frac{x}{2(n+1)}[J_n(x) + J_{n+2}(x)]$$

Prob. 2.12 From the generating function for $J_n(x)$,

$$e^{\rho(t-1/t)/2} = \sum_{n=-\infty}^{\infty} t^n J_n(\rho)$$

If $t = e^{j\phi}$,

$$\frac{1}{2}\left(t - \frac{1}{t}\right) = \frac{1}{2}(e^{j\phi} - e^{-j\phi}) = j \sin\phi$$

Hence

$$e^{j\rho \sin\phi} = \sum_{n=-\infty}^{\infty} (e^{j\phi})^n J_n(\rho)$$

If $t = e^{-j\phi}$,

$$\frac{1}{2}\left(t - \frac{1}{t}\right) = -j \sin\phi$$

$$e^{-j\rho \sin\phi} = \sum_{n=-\infty}^{\infty} (e^{j\phi})^n J_n(-\rho)$$

But $J_n(-\rho) = (-1)^n J_n(\rho)$. Thus

$$e^{\pm j\rho \sin\phi} = \sum_{n=-\infty}^{\infty} (\pm)^n e^{jn\phi} J_n(\rho)$$

Prob. 2.13 From Prob. 2.12,

$$e^{j\rho \sin\phi} = \sum_{n=-\infty}^{\infty} e^{jn\phi} J_n(\rho)$$

Equating real and imaginary parts,

$$\cos(\rho \sin\phi) = \sum_{n=-\infty}^{\infty} J_n(\rho) \cos n\phi$$

$$\sin(\rho \sin\phi) = \sum_{n=-\infty}^{\infty} J_n(\rho) \sin n\phi$$

But $J_{-n}(\rho) = (-1)^n J_n(\rho)$, $\cos(-n\phi) = \cos n\phi$, $\sin(-n\phi) = -\sin n\phi$. Hence

$$\cos(\rho \sin\phi) = J_0(\rho) + 2\sum_{n=1}^{\infty} J_{2n}(\rho) \cos 2n\phi$$

$$\sin(\rho \sin \phi) = 2 \sum_{n=1}^{\infty} J_{2n+1}(\rho) \sin(2n-1)\phi$$

By setting $\phi = \pi/2$ and replacing ρ with x, we obtain

$$\cos x = J_o(x) + 2 \sum_{n=1}^{\infty} (-1)^n J_{2n}(x)$$

and

$$\sin x = 2 \sum_{n=1}^{\infty} (-1)^{n+1} J_{2n+1}(x)$$

Prob. 2.14 (a)

$$J_{1/2}(x) = \sum_{k=0}^{\infty} \frac{(-1)^k (x/2)^{2k+1/2}}{k!\Gamma(k+3/2)}$$

$$= \frac{(x/2)^{1/2}}{\sqrt{\pi}/2} - \frac{(x/2)^{5/2}}{\frac{3}{2}\sqrt{\pi}/2} + \frac{(x/2)^{7/2}}{2\frac{5}{2}\frac{5}{2}\sqrt{\pi}/2} - \cdots$$

$$= \frac{(x/2)^{1/2}}{\sqrt{\pi}/2} \left[1 - \frac{x^2}{3!} + \frac{x^4}{5!} - \cdots \right]$$

$$= \sqrt{\frac{2}{\pi x}} \sin x$$

(b)

$$J_{-1/2}(x) = \sum_{k=0}^{\infty} \frac{(-1)^k (x/2)^{2k-1/2}}{k!\Gamma(k+1/2)}$$

$$= \frac{(x/2)^{-1/2}}{\sqrt{\pi}} \left[1 - \frac{x^2}{2!} + \frac{x^4}{4!} - \cdots \right]$$

$$= \sqrt{\frac{2}{\pi x}} \cos x$$

(c)

$$\frac{d}{dx}(x^{-n} J_n(x)) = \frac{d}{dx} \sum_{m=0}^{\infty} \frac{(-1)^m x^{2m}}{2^{n+2m} m! \Gamma(n+m+1)}$$

$$= \sum_{m=0}^{\infty} \frac{2m(-1)^m x^{2m-1}}{2^{n+2m} m! \Gamma(n+m+1)}$$

$$= x^{-n} \sum_{m=0}^{\infty} \frac{(-1)^m x^{2m+n-1}}{2^{n+2m-1}(m-1)! \Gamma(n+m+1)}$$

Replacing $(m-1)$ with k gives

$$\frac{d}{dx}(x^{-n}J_n(x)) = -x^{-n}\sum_{k=0}^{\infty}\frac{(-1)^k x^{n+1+2k}}{2^{n+1+2k}k!\Gamma(n+m+2)}$$

$$= -x^{-n}J_{n+1}(x)$$

(d) Since

$$J_n(x) = \sum_{k=0}^{\infty}\frac{(-1)^k x^{2k+n}}{2^{2k+n}k!(n+k)!}$$

$$J_n'(x) = \sum_{k=0}^{\infty}\frac{(-1)^k(2k+n)x^{2k+n-1}}{2^{2k+n}k!(n+k)!}$$

$$\frac{d^n}{dx^n}J_n(x) = \sum_{k=0}^{\infty}\frac{(-1)^k(2k+n)!x^{2k}}{2^{2k+n}k!(2k)!(n+k)!}$$

$$= \frac{n!}{2^n n!}x^0 - \frac{(n+2)!x^2}{2^{n+2}2!4!(n+1)!} + \cdots$$

$$\left.\frac{d^n}{dx^n}J_n(x)\right|_{x=0} = \frac{1}{2^n}$$

(e)

$$\frac{d}{dx}[xz_n(x)] = z_n(x) + x\frac{d}{dx}z_n(x)$$

$$= z_n + \frac{x}{2n+1}[nz_{n-1} - (n+1)z_{n+1}(x)]$$

$$= z_n + \frac{xn}{2n+1}z_{n-1} - \frac{x(n+1)}{2n+1}\Big[\frac{(2n+1)}{x}z_n - z_{n-1}\Big]$$

$$= -nz_n(x) - xz_{n-1}(x)$$

$$= -nz_n(x) - x\Big[\frac{2n+1}{x}z_n - z_{n+1}\Big]$$

$$= (n+1)z_n(x) - xz_{n+1}(x)$$

Prob. 2.15

$$\frac{\partial I_o}{\partial a} = -\int_0^{\infty}\lambda e^{-\lambda a}J_o(\lambda\rho)d\lambda = -I_1$$

Hence

$$I_1 = -\frac{\partial I_o}{\partial a} = -\frac{\partial}{\partial a}\frac{1}{(\rho^2+a^2)^{1/2}}$$

$$= \frac{a}{(\rho^2+a^2)^{3/2}}$$

Similarly,

$$\frac{\partial^2 I_o}{\partial a^2} = \int_0^\infty \lambda^2 e^{-\lambda a} J_o(\lambda \rho) d\lambda = I_2$$

Hence

$$I_2 = \frac{\partial^2}{\partial a^2} \frac{1}{(\rho^2 + a^2)^{1/2}} = \frac{2a^2 - \rho^2}{(\rho^2 + a^2)^{5/2}}$$

Prob. 2.16 The result is shown in the table below.

m	$J_0(x)$	$J_1(x)$	$J_2(x)$	$J_3(x)$	$J_4(x)$	$J_5(x)$
1	2.4048	3.8317	5.1356	6.3802	7.5883	8.7715
2	5.5201	7.0156	8.4172	9.7610	11.06747	12.3386
3	8.6537	10.1735	11.6198	13.0152	14.3725	15.7002
4	11.7915	13.3237	14.7960	16.2235	17.6160	18.9801
5	14.9309	16.4706	17.9598	19.4095	20.8769	22.2178

Prob. 2.17(a)

$$\int_{-1}^1 P_n(x) P_m(x) dx = \begin{cases} 0, & n \neq m \\ \frac{2}{2n+1}, & n = m \end{cases}$$

Hence

$$\int_{-1}^1 P_1(x) P_2(x) dx = 0$$

(b)

$$\int_{-1}^1 [P_4(x)]^2) dx = \frac{2}{8+1} = \frac{2}{9}$$

(c)

$$\int_0^1 x^2 P_3(x) dx = \frac{1}{2} \int_0^1 (5x^5 - 3x^3) dx = \frac{1}{24}$$

Prob. 2.18

$$f(x) = \int_{n=0}^\infty a_n P_n(x)$$

where

$$a_n = \frac{2n+1}{2} \int_{-1}^1 f(x) P_n(x) dx$$

(a) If

$$f(x) = \begin{cases} 0, & -1 < x < 0 \\ 1, & 0 < x < 1 \end{cases}$$

$$a_n = \frac{2n+1}{2} \int_0^1 P_n(x)\,dx$$

But

$$\int P_n(x)\,dx = \frac{P_{n+1}(x) - P_{n-1}(x)}{2n+1}$$

$$a_n = \frac{1}{2}[P_{n+1}(x) - P_{n-1}(x)\Big|_0^1$$

$$= \frac{1}{2}\Big[P_{n+1}(1) - P_{n-1}(1) - P_{n+1}(0) + P_{n-1}(0)\Big]$$

$$= \frac{1}{2}[P_{n-1}(0) - P_{n+1}(0)]$$

But

$$P_n(0) = \begin{cases} 0, & n = \text{odd} \\ (-1)^{n/2}\frac{1\cdot 3\cdot 5\cdots(n-1)}{2\cdot 4\cdot 6\cdots n}, & n = \text{even} \end{cases}$$

It is evident that $a_n = 0$ when $n =$ even.

$$a_0 = \frac{1}{2}, \quad a_1 = \frac{7}{4}, \quad a_3 = -\frac{7}{16}, \quad a_5 = \frac{18}{16}, \text{ etc.}$$

Hence

$$f(x) = \frac{1}{2}\sum_{n=0}^{\infty}[P_{n-1}(0) - P_{n+1}(0)]P_n(x)$$

$$= \frac{1}{2} + \frac{3}{4}P_1 - \frac{7}{16}P_3 + \cdots$$

(b)

$$f(x) = x^3 = A_0 P_0(x) + A_1 P_1(x) + A_2 P_2(x) + \cdots$$

$$= A_0(1) + A_1(x) + A_2(\frac{3x^2 - 1}{2}) + A_3(\frac{5x^3 - 3x}{2}) + \cdots$$

Since $f(x)$ is of degree 3, $A_n = 0$ for $n \geq 4$. Equating coefficients gives

constant: $0 = A_0 - A_3/2 \quad \rightarrow A_2 = 2A_0$

x: $0 = A_1 - 3A_3/2 \quad \rightarrow A_1 = 3A_0/2$

x^2: $0 = A_2 \quad \rightarrow A_2 = 0 = A_0$

x^3: $1 = 5A_3/2 \quad \rightarrow A_3 = 2/5, \; A_1 = 3/5$ Hence

$$f(x) = \frac{3}{5}P_1(x) + \frac{2}{5}P_3(x)$$

(c) If

$$f(x) = \begin{cases} 0, & -1 < x < 0 \\ x, & 0 < x < 1 \end{cases} = \sum_{n=1}^{\infty} A_n P_n(x)$$

$$A_n = \frac{2n+1}{2} \int_0^1 x P_n(x)\,dx$$

$$A_0 = \frac{1}{2} \int_0^1 x(1)\,dx = \frac{1}{4}$$

$$A_1 = \frac{3}{2} \int_0^1 x(x)\,dx = \frac{1}{2}$$

$$A_2 = \frac{5}{2} \int_0^1 \frac{1}{2}(3x^3 - x)\,dx = \frac{5}{16}$$

$$A_3 = \frac{7}{2} \int_0^1 \frac{1}{2}(5x^4 - 3x^2)\,dx = 0$$

$$A_4 = \frac{9}{2} \int_0^1 \frac{1}{8}(35x^5 - 30x^3 + 3x)\,dx = -\frac{3}{32}$$

$$A_5 = \frac{11}{2} \int_0^1 \frac{1}{8}(63x^6 - 70x^4 + 15x^2)\,dx = 0, \text{etc}$$

$$f(x) = \frac{1}{4}P_0(x) + \frac{1}{2}P_1(x) + \frac{5}{16}P_2(x) - \frac{3}{32}P_4(x) + \cdots$$

(d)

$$f(x) = \begin{cases} 1 + x, & -1 < x < 0 \\ 1 - x, & 0 < x < 1 \end{cases}$$

$f(x)$ is an even function; $A_n = 0 = 0$ if $n =$ odd. For even values of n,

$$A_n = \frac{2n+1}{2} \int_{-1}^1 f(x) P_n(x)\,dx$$

or

$$A_n \cdot \frac{2}{2n+1} = \int_{-1}^0 (1+x) P_n(x)\,dx + \int_0^1 (1-x) P_n(x)\,dx$$

$$= \int_{-1}^1 P_n(x)\,dx - \int_0^{-1} x P_n(x)\,dx - \int_0^1 x P_n(x)\,dx$$

For $n = 0$,

$$2A_0 = \int_{-1}^1 (1)\,dx - \int_0^{-1} x\,dx - \int_0^1 x\,dx = 1 \quad \rightarrow \quad A_0 = \frac{1}{2}$$

For $n = 2$,

$$\frac{2}{5}A_2 = -\frac{1}{4} \quad \rightarrow \quad A_2 = -\frac{5}{8}, \text{ etc.}$$

$$f(x) = \frac{1}{2} - \frac{5}{8}P_2(x) + \frac{3}{16}P_4(x) + \cdots$$

Prob. 2.19(a) Let $U(r,\theta) = R(r)H(\theta)$. Upon separating the variables,

$$U_n(r,\theta) = \left(A_n r^n + B_n r^{-(n+1)}\right)P_n(\cos\theta)$$

$B_n = 0$ since U is finite at $r = 0$.

$$U(r,\theta) = \sum_{n=0}^{\infty} A_n r^n P_n(\cos\theta)$$

$$U(a,\theta) = \sum_{n=0}^{\infty} A_n a^n P_n(\cos\theta) = \begin{cases} 1, & 0 < \theta < \pi/2 \\ 0, & \text{otherwise} \end{cases}$$

$$A_n = a^{-n}\frac{2n+1}{2}\int_0^{\pi} U(a,\theta)P_n(\cos\theta)\sin\theta d\theta$$

$$= a^{-n}\frac{2n+1}{2}\int_0^{\pi/2} P_n(\cos\theta)\sin\theta d\theta$$

$$A_0 = \frac{1}{2}, \qquad A_n = a^{-n}\frac{(2n+1)}{2n(n+1)}P_n'(0), \ n \neq 0$$

Thus,

$$U(r,\theta) = \frac{1}{2} + \sum_{n=1}^{\infty}\frac{(2n+1)}{2n(n+1)}P_n'(0)(r/a)^n P_n(\cos\theta)$$

(b) As in part (a),

$$U_n(r,\theta) = \left(A_n r^n + \frac{B_n}{r^{n+1}}\right)P_n(\cos\theta)$$

$A_n = 0$ since U is finite as $r \rightarrow \infty$.

$$U_n = \sum_{n=0}^{\infty}\frac{B_n}{r^{n+1}}P_n(\cos\theta)$$

$$\left.\frac{dU_n}{dr}\right|_{r=a} = \cos\theta + 3\cos^3\theta = \sum_{n=0}^{\infty} -\frac{B_n(n+1)}{a^{n+2}}P_n(\cos\theta)$$

$$B_n = -\frac{a^{n+2}}{n+1}\frac{2n+1}{2}\int_0^\pi (\cos\theta + 3\cos^3\theta)P_n(\cos\theta)\sin\theta d\theta$$

But

$$\cos\theta + 3\cos^3\theta = \frac{6}{5}P_3 + \frac{14}{3}P_1$$

Hence

$$B_n = -\frac{a^{n+2}}{n+1}\frac{2n+1}{2}\int_{-1}^1 \left[\frac{6}{5}P_3 + \frac{14}{5}P_1\right]P_n(x)dx$$

From this,

$$B_1 = -\frac{7}{5}a^3, \; B_3 = -\frac{3}{10}a^5, \qquad B_n = 0, \; n \neq 1,3$$

$$U(r,\theta) = -\frac{7}{5}\frac{a^3}{r^2}P_1(\cos\theta) - \frac{3}{10}\frac{a^5}{r^4}P_3(\cos\theta)$$

(c)

$$U(r,\theta) = \sum_{n=0}^\infty A_n r^n P_n(x), \; x = \cos\theta$$

$$U(a,\theta) = \sin^2\theta = 1 - x^2 = \frac{2}{3}P_0 - \frac{2}{3}P_2 = \sum_{n=0}^\infty A_n a^n P_n$$

Equating both sides yields

$$A_0 = \frac{2}{3}, \; A_1 = 0, \; A_2 = -\frac{2}{3a^2}, \; A_n = 0, \; n \geq 3$$

Hence

$$U(r,\theta) = \frac{2}{3}\left[1 - \frac{r^2}{a^2}P_3(\cos\theta)\right]$$

Prob. 2.20 From the previous problem,

$$V_n(r,\theta) = \left(A_n r^n + \frac{B_n}{r^{n+1}}\right)P_n(x), \; x = \cos\theta$$

subject to the following boundary conditions:

$$V(a,\theta) = \begin{cases} V_o, & 0 < \theta < \pi/2, \; 0 < x < 1 \\ 0, & \pi/2 < \theta < \pi, \; -1 < x < 0 \end{cases}$$

V is bounded at $r = 0, \infty$.

(a) Inside the sphere, $r < a$, $B_n = 0$

$$V(r,\theta) = \sum_{n=0}^\infty A_n r^n P_n(\cos\theta)$$

$$V(a, \theta) = \sum_{n=1}^{\infty} A_n a^n P_n(x)$$

$$A_n = a^{-n} \frac{2n+1}{2} \int_{-1}^{1} V(a, x) P_n(x) dx$$

$$= \frac{V_o}{2a^n} [P_{n-1}(0) - P_{n+1}(0)]$$

$$A_0 = \frac{V_o}{2}, \quad A_1 = \frac{3V_o}{4a}, \quad A_3 = -\frac{7V_o}{16a}, \quad A_2 = 0 = A_4, \quad A_5 = \frac{11V_o}{32a^5}, \text{ etc.}$$

$$V(r, \theta) = \frac{V_o}{2} \left[1 + \frac{3r}{2a} P_1(\cos\theta) - \frac{7r^3}{8a^3} P_3(\cos\theta) + \cdots \right]$$

(b) Outside the sphere, $r > a$, $A_n = 0$.

$$V(r, \theta) = \sum_{n=0}^{\infty} \frac{B_n}{r^{n+1}} \Big) P_n(x), \quad x = \cos\theta$$

$$V(a, \theta) = \sum_{n=0}^{\infty} \frac{B_n}{a^{n+1}} \Big) P_n(x)$$

B_n are obtained just as in (a).

$$V(r, \theta) = \frac{V_o}{2r} \left[1 + \frac{3a}{2r} P_1(\cos\theta) - \frac{7a^3}{8r^3} P_3(\cos\theta) + \frac{11a^5}{16r^5} P_5(\cos\theta) + \cdots \right]$$

Prob. 2.21

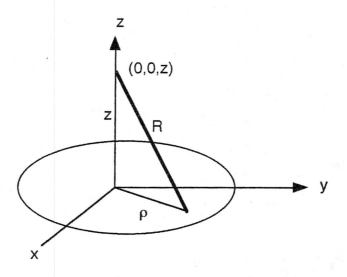

$$V = \int_S \frac{\rho_s dS}{4\pi\epsilon R} = \rho_o \int \int \frac{\rho d\phi d\rho}{4\pi\epsilon(z^2 + \rho^2)^{1/2}}$$

$$= \frac{\rho_o}{4\pi\rho}(2\pi)(z^2 + \rho^2)^{1/2}\Big|_0^a$$

$$= \frac{\rho_o}{2\epsilon}\left[(z^2 + a^2)^{1/2} - z\right]$$

To find V at any point, we follow the same procedure as in Example 2.6(b).

$$V(r, \theta) = \sum_{n=0}^{\infty}\left(A_n r^n + \frac{B_n}{r^{n+1}}\right)P_n(\cos\theta)$$

For $r \leq a$, $B_n = 0$ and for $r \geq a$, $A_n = 0$. Hence

$$V(r, \theta) = \begin{cases} \displaystyle\sum_{n=0}^{\infty} A_n r^n P_n(\cos\theta), & r \leq a \\ \displaystyle\sum_{n=0}^{\infty} \frac{B_n}{r^{n+1}} P_n(\cos\theta), & r \geq a \end{cases}$$

As in Example 2.6(b), the values of A_n and B_n are determined by noting that at $\theta = 0$, $P_n(\cos\theta) = 1$, $r = z$, and

$$V(0, 0, z) = \frac{\rho_o}{2\epsilon}\left[(z^2 + a^2)^{1/2} - z\right]$$

Thus,

$$V(r,\theta) = \begin{cases} \dfrac{\rho_o a}{2\epsilon} \displaystyle\sum_{n=0}^{\infty} \dfrac{(-1)^n (2n)!}{(n!2^n)^2}(r/a)^{2n}P_{2n} - \dfrac{\rho_o z}{2\epsilon}, & r \leq a \\[3mm] \dfrac{\rho_o a}{2\epsilon} \displaystyle\sum_{n=0}^{\infty} \dfrac{(-1)^{n+1}(2n+2)!}{[(n+1)!2^{n+1}]^2}(a/r)^{2n+1}P_{2n}, & r > a \end{cases}$$

Prob. 2.22(a)

$$G(x,t) = (1 - 2xt + t^2)^{-1/2} = \sum_{n=0}^{\infty} t^n P_n(x) \tag{1}$$

$$\frac{dG}{dt} = \frac{x-t}{(1-2xt+t^2)^{3/2}} = \sum_{n=0}^{\infty} t^n P_n(x)t^{n-1} \tag{2}$$

Substituting (1) into (2) and rearranging terms,

$$(1 - 2xt + t^2)\sum_{n=0}^{\infty} t^n P_n(x)t^{n-1} + (t - x)\sum_{n=0}^{\infty} t^n P_n(x)t^n = 0$$

By separating individual summations and using distinct indices, we get

$$\sum_{m=0}^{\infty} mP_m(x)t^{m-1} - \sum_{n=0}^{\infty} 2nxP_n(x)t^n + \sum_{s=0}^{\infty} sP_s(x)t^{s+1} + \sum_{s=0}^{\infty} P_s(x)t^{s+1} - \sum_{n=0}^{\infty} xP_n(x)t^n = 0$$

Letting $m = n + 1$ and $s = n - 1$, we obtain

$$(2n + 1)xP_n(x) = (n + 1)P_{n+1}(x) + nP_{n-1}(x)$$

as required.

(b) For P_6, let $n = 5$

$$6P_6(x) = 11xP_5(x) - 5P_4(x)$$

which leads to

$$P_6(x) = \frac{1}{16}(231x^6 - 315x^4 + 105x^2 - 5)$$

Simiarly, for P_7, let $n = 6$

$$7P_7(x) = 13xP_6(x) - 6P_5(x)$$

which leads to

$$P_7(x) = \frac{1}{16}(429x^7 - 693x^5 + 315x^5 - 35x)$$

Prob. 2.23(a) First, we prove that $\int_{-1}^{1} P_n(x)P_m(x)dx = 0$, $m \neq n$. $P_n(x)$ satisfy Legendre's differential equation:

$$\frac{d}{dx}[(1-x^2)P_n'(x)] + n(n+1)P_n(x) = 0$$

Multiplying by $P_m(x)$ and integrating results in

$$\int_{-1}^{1} P_m(x)\frac{d}{dx}[(1-x^2)P_n'(x)]dx + n(n+1)\int_{-1}^{1} P_m(x)P_n(x)dx = 0$$

We integrate the first term by parts with $u = P_m(x)$, $du = P_m'(x)dx$ so that

$$\int_{-1}^{1} P_m(x)\frac{d}{dx}[(1-x^2)P_n'(x)]dx = P_m(x)P_n'(x)(1-x^2)\Big|_{-1}^{1} - \int_{-1}^{1}(1-x^2)P_n'(x)P_m'(x)dx$$

where the first term on the right hand side becomes zero. Thus

$$-\int_{-1}^{1}(1-x^2)P_n'(x)P_m'(x)dx + n(n+1)\int_{-1}^{1} P_m(x)P_n(x)dx = 0 \qquad (1)$$

Interchanging m and n,

$$-\int_{-1}^{1}(1-x^2)P_m'(x)P_n'(x)dx + m(m+1)\int_{-1}^{1} P_n(x)P_m(x)dx = 0 \qquad (2)$$

Subtracting (2) from (1),

$$(n-m)(n+m+1)\int_{-1}^{1} P_m(x)P_n(x)dx = 0$$

Since m and n are positive integers, if $m \neq n$,

$$\int_{-1}^{1} P_m(x)P_n(x)dx = 0$$

For case $m = n$, we use Rodrigues formula and integrate by parts n times.

$$\int_{-1}^{1} P_n(x)dx = \frac{1}{2^{2n}(n!)^2}\int_{-1}^{1}\left[\frac{d^n}{dx^n}(1-x^2)^n\right]^2 dx$$

$$= \frac{1}{2^{2n}(n!)^2}(2n)!\int_{-1}^{1}(1-x^2)^n dx$$

$$= \frac{2}{2n+1}$$

Hence

$$\int_{-1}^{1} P_m(x)P_n(x)dx = \frac{2}{2n+1}\delta_{mn}$$

Similarly,

$$\int_{-1}^{1} [P_n^m(x)]^2 dx = \int_{-1}^{1}\left[(1-x^2)^{m/2}\frac{d^m P_n}{dx^m}\right]^2 dx$$

$$= (-1)^m \int_{-1}^{1} P_n \frac{d^m}{dx^m}\left[(1-x^2)^m \frac{d^m P_n}{dx^m}\right]dx$$

$$= \frac{(n+m)!}{n-m)!}\int_{-1}^{1} P_n c_n x^n dx$$

where c_n is the coefficient of x^n in $P_n(x)$. Since the P_n are orthogonal, each $P_n(x)$ must therefore be orthogonal to all powers of x less than n, i.e.

$$\int_{-1}^{1} x^k P_n(x)dx = 0, \quad k = 0,1,2,\cdots,n-1$$

Thus

$$\int_{-1}^{1} [P_n^m(x)]^2 dx = \frac{(n+m)!}{n-m)!}\int_{-1}^{1} P_n dx$$

$$= \frac{2}{2n+1}\frac{(n+m)!}{(n-m)!}$$

We can follow the same steps in (a) to show that

$$\int_{-1}^{1} P_n^m(x)P_k^m(x)dx = 0, \quad n \neq k$$

Hence

$$\int_{-1}^{1} P_n^m(x)P_k^m(x)dx = \frac{2}{2n+1}\frac{(n+m)!}{(n-m)!}\delta_{nk}$$

Prob. 2.24 Assume the solution

$$V(r,\theta) = \sum_{n=0}^{\infty}\left(A_n r^n + \frac{B_n}{r^{n+1}}\right)P_n(\cos\theta)$$

$$= A_o + A_1 r\cos\theta + A_2 r^2 P_2(\cos\theta) + \cdots$$

$$+ \frac{B_o}{r} + \frac{B_1}{r^2}\cos\theta + \frac{B_2}{r^3}P_2(\cos\theta) + \cdots$$

The second condition

$$V(r\to\infty) = V_o + E_o r\cos\theta$$

is satisfied if $A_o = V_o$ and $A_1 = E_o$. Imposing the first condition, $V(r, a) = V_o$, results in

$$V = V_o + E_o a \cos\theta + A_2 a^2 P_2(\cos\theta) + \cdots$$
$$+ \frac{B_o}{a} + \frac{B_1}{a^2}\cos\theta + \frac{B_2}{a^3}P_2(\cos\theta) + \cdots$$

Hence

$$B_o = 0, \; B_1 = -E_o a^3, \; A_2 = A_3 = \cdots = B_2 = B_3 = \cdots = 0$$

and

$$V(r,\theta) = V_o + E_o\left(1 - \frac{a^3}{r^3}\right) r\cos\theta$$

$$\mathbf{E} = -\nabla V = -\frac{\partial V}{\partial r}\mathbf{a}_r - \frac{1}{r}\frac{\partial V}{\partial\theta}\mathbf{a}_\theta$$
$$= -E_o\left(1 + \frac{2a^3}{r^3}\right)\cos\mathbf{a}_r + E_o\left(1 - \frac{a^3}{r^3}\right)\sin\theta\mathbf{a}_\theta$$

$E_{\max}$ occurs when $\to a$, i.e.

$$\lim_{r\to a}|\mathbf{E}| = 3E_o\cos\theta$$

or

$$|\mathbf{E}|_{\max} = 3E_o$$

Prob. 2.25 Since the boundary condition does not depend on ϕ, $m = 0$. So we look for the solution of the form

$$V(r,\theta) = \sum_{n=0}^{\infty} A_n r^n P_n(\cos\theta)$$

The boundary conditions can be written as

$$V(a,\theta) = 2P_0(\cos\theta) + 3P_1(\cos\theta) + 2P_2(\cos\theta)$$

Hence the required solution is

$$V(r,\theta) = 2(r/a)^2 P_2(\cos\theta) + 3(r/a)P_1(\cos\theta) + 2P_0(\cos\theta)$$

Prob. 2.26 Since V is inside the sphere, the solution assumes the form

$$V(r,\theta) = \sum_{n=1}^{\infty} \frac{A_n}{r^{(n+1)}}P_n$$

$$\frac{\partial V}{\partial r} = -\sum_{n=1}^{\infty} \frac{A_n(n+1)}{r^{n+2}} P_n$$

$$\frac{\partial V}{\partial r}(a,0) = -\sum_{n=1}^{\infty} \frac{A_n(n+1)}{a^{n+2}} P_n = \cos\theta + 3\cos 3\theta$$

which contains $P_1 = \cos\theta$ and $P_3 = \frac{5}{2}\cos^3\theta - \frac{3}{2}\cos\theta$.

$$\cos\theta + 3\cos 3\theta = \frac{5}{2}P_1 + \frac{6}{5}P_3$$

Thus

$$-\frac{2A_1 P_1}{a^3} - \frac{4A_3 P_3}{a^5} = \frac{5}{2}P_1 + \frac{6}{5}P_3$$

which implies that $A_1 = -\frac{5a^3}{4}$, $A_3 = -\frac{6a^5}{20}$ Thus

$$V(r,\theta) = -\frac{7}{5}\frac{a^3}{r^2}P_1(\cos\theta) - \frac{3}{10}\frac{a^5}{r^4}P_3(\cos\theta)$$

Prob. 2.27 Let

$$V(r,\theta) = \sum_{n=0}^{\infty}(a_n r^n + b_n r^{-n-1})P_n(\cos\theta)$$

Since V_1 is finite at the origin,

$$V_1 = \sum_{n=0}^{\infty} a_n r^n P_n(\cos\theta), \ r \leq a$$

and since $V_2 = E_o r \cos\theta$ as $r \to \infty$,

$$V_2 = -E_o r \cos\theta + \sum_{n=0}^{\infty} b_n r^{-n-1} P_n(\cos\theta), \ r \geq a$$

Imposing the boundary conditions,

$$V_1 = V_2 \qquad \to \qquad a_1 = -E_o + b_1/a^3$$

$$\epsilon_r \frac{\partial V_1}{\partial r} = \frac{\partial V_2}{\partial r} \qquad \to \qquad \epsilon_r a = -E_o - 2b_1/a^3$$

Hence

$$a_1 = -\frac{3E_o}{\epsilon_r + 2}, \qquad b_1 = E_o a^3 \frac{(\epsilon_r - 1)}{\epsilon_r + 2}$$

and

$$V_1(r, \theta) = -\frac{3E_o}{\epsilon_r + 2} \cos, \ r \le a$$

$$V_2(r, \theta) = -E_o r \cos \theta + E_o \frac{a^3}{r^2} \frac{(\epsilon_r - 1)}{\epsilon_r + 2} \cos \theta$$

Prob. 2.28 (a) $P_n^m(x)$ should satisfy

$$P_n^m(x) = (1 - x^2)^{m/2} \frac{d^m}{dx^m} P_n(x) \tag{1}$$

and

$$(1 - x^2)\frac{d^2 P_n^m}{dx^2} - 2x\frac{dP_n^m}{dx} \left[n(n+1) - \frac{m^2}{1 - x^2} \right] P_n^m(x) = 0 \tag{2}$$

From (1),

$$\frac{dP_n^m}{dx} = \frac{m}{2}(-2x)(1 - x^2)^{m/2-1}\frac{d^m}{dx^m} P_n(x) + (1 - x^2)^{m/2}\frac{d^{m+1}}{dx^{m+1}} P_n(x) \tag{3}$$

or

$$\frac{dP_n^m}{dx} = -mx(1 - x^2)^{m/2-1} P_n^m + (1 - x^2)^{m/2} P_n^{m+1} \tag{4}$$

From (3),

$$\frac{d^2 P_n^m}{dx^2} = -\frac{mP_n^m}{1 - x^2} + \frac{m(m-2)x^2 P_n^m}{(1 - x^2)^2} - \frac{2mx P_n^{m+1}}{(1 - x^2)^{3/2}} + \frac{P_n^{m+2}}{1 - x^2} \tag{5}$$

Substituting (4) and (5) into (2) gives

$$0 = -mP_n^m + \frac{m(m-2)x^2}{1 - x^2} P_n^m - \frac{2mx}{(1 - x^2)^{1/2}} + P_n^{m+1} + P_n^{m+2}$$

$$- \frac{2mx^2}{1 - x^2} P_n^m - \frac{2x}{(1 - x^2)^{1/2}} P_n^{m+1} + [n(n+1) - \frac{m^2}{1 - x^2}] P_n^m$$

Replacing every m with $m - 1$, we obtain

$$P_n^{m+1} = \frac{2xm}{(1 - x^2)^{3/2}} P_n^m - [n(n+1) - m(m-1)] P_n^{m-1}$$

as required.

(b) Since $P_n^m(x) = (1 - x^2)^{m/2}\frac{d^m P_n}{dx^m}$,

$$P_3^2(x) = (1 - x^2)\frac{d^2 P_3}{dx^2}$$

$$= (1 - x^2)\frac{d^2}{dx^2}\frac{1}{2}(5x^3 - 3x)$$

$$= 15x(1 - x^2)$$

$$P_3^3(x) = (1-x^2)^{3/2}\frac{d^3 P_3}{dx^3}$$

$$= \frac{1}{2}(1-x^2)^{3/2}\frac{d}{dx}(30x)$$

$$= 15(1-x^2)^{3/2}$$

$$P_4^1(x) = (1-x^2)^{1/2}\frac{dP_4}{dx}$$

$$= \frac{1}{8}(1-x^2)^{1/2}\frac{d}{dx}(35x^4 - 30x^2 + 3)$$

$$= \frac{5}{2}(7x^3 - 3x)(1-x^2)^{3/2}$$

$$P_4^2(x) = (1-x^2)^2\frac{d^2 P_4}{dx^2}$$

$$= \frac{1}{8}(1-x^2)\frac{d}{dx}(140x^3 - 60x)$$

$$= \frac{15}{2}(7x^2 - 1)(1-x^2)$$

Prob. 2.29 Let

$$V = cos2\phi \sin^2\theta = \int_{n=0}^{\infty}\int_{m=0}^{n}(a_{mn}\cos m\phi + b_{mn}\sin m\phi)P_n^m(\cos\theta)$$

Since $\sin m\phi$ and $\cos m\phi$ are orthogonal functions, $a_{mn} = 0 = b_{mn}$ except a_{n2}. Hence

$$V = \cos 2\phi \sin^2\theta = \cos 2\phi \int_{n=0}^{\infty} a_{2n}P_n^2(\cos\theta)$$

or

$$\sin^2\theta = 1 - x^2 = \int_{n=0}^{\infty} a_{2n}P_n^2(\cos\theta)$$

Using orthogonality property,

$$\frac{2}{2n+1}\frac{n+2}{(n-2)!}a_{2n} = \int_{-1}^{2}(1-x^2)P_n^2(x)dx$$

$$= \int_{-1}^{1}\frac{1}{3}P_2^2(x)P_n^2(x)dx$$

$$= \begin{cases} 0, & n \neq 2 \\ \frac{1}{3}\frac{2(n+2)}{(2n+1)(n-2)!}, & n = 2 \end{cases}$$

i.e. $a_{2n} = \dfrac{1}{3}$ and thus

$$V = \cos 2\phi \sin^2\theta = \frac{1}{3}\cos 2\phi P_2^2(\cos\theta)$$

Prob. 2.30 In prolate spheroidal coordinates (ζ, η, ϕ),

$$x = d\sinh u \sin v \cos\phi, \ \ y = d\sinh u \sin v \sin\phi, \ \ z = d\cosh u \cos v,$$
$$\zeta = \cosh u, \ 1 \le \zeta \le \infty, \quad \eta = \cos v, \ -1 \le \eta \le 1$$

The wave equation $\nabla^2\Phi + k^2\Phi = 0$ can be represented as

$$\frac{1}{d^2(\sinh^2 u + \sin^2 v)\sinh u}\frac{\partial}{\partial u}\left(\sinh u \frac{\partial\Phi}{\partial u}\right)$$
$$+\frac{1}{d^2(\sinh^2 u + \sin^2 v)\sin v}\frac{\partial}{\partial v}\left(\sinh v \frac{\partial\Phi}{\partial v}\right)$$
$$+\frac{1}{d^2\sinh^2 u \sin^2 v}\frac{\partial\Phi}{\partial\phi^2} + k^2\Phi = 0$$

Multiplying by $d^2(\sinh^2 u + \sin^2 v)$ gives

$$\frac{1}{\sinh u}\frac{\partial}{\partial u}\left(\sinh u \frac{\partial\Phi}{\partial u}\right) + \frac{1}{\sin v}\frac{\partial}{\partial v}\left(\sinh v \frac{\partial\Phi}{\partial v}\right)$$
$$+\frac{(\sinh^2 u + \sin^2 v)}{\sinh^2 u \sin^2 v}\frac{\partial\Phi}{\partial\phi^2} + k^2\Phi(\sinh^2 u + \sin^2 v) = 0$$

Since $\zeta = \cosh u$ and $\eta = \cos v$,

$$\sin v = \sqrt{1-\eta^2}, \qquad \sinh u = \sqrt{\zeta^2 - 1}$$
$$\frac{\sinh^2 u + \sin^2 v}{\sinh^2 u \sin^2 v} = \left[\frac{1}{\zeta^2-1} + \frac{1}{1-\eta^2}\right]$$
$$\frac{d}{du} = \sinh u \frac{d}{d\zeta} \ \rightarrow \ \frac{d}{d\zeta} = \frac{1}{\sinh u}\frac{d}{du}$$
$$\frac{d}{dv} = -\sin v \frac{d}{d\eta} \ \rightarrow \ \frac{d}{d\eta} = -\frac{1}{\sin v}\frac{d}{dv}$$

Substituting all these,

$$\frac{\partial}{\partial\zeta}\left[(\zeta^2-1)\frac{\partial\Phi}{\partial\zeta}\right] + \frac{\partial}{\partial\eta}\left[(1-\eta^2)\frac{\partial\Phi}{\partial\eta}\right]$$
$$+\left[\frac{1}{\zeta^2-1} + \frac{1}{1-\eta^2}\right]\frac{\partial^2\Phi}{\partial\phi^2} + k^2 d^2(\zeta^2-\eta^2)\Phi = 0$$

Substituting $\Phi = \Psi_1(\zeta)\Psi_2(\eta)\Psi_3(\phi)$ and dividing through by $\Psi_1\Psi_2\Psi_3$, we obtain

$$\frac{1}{\Psi_1}\frac{\partial}{\partial\zeta}\left[(\zeta^2-1)\frac{\partial\Psi_1}{\partial\zeta}\right] + \frac{1}{\Psi_2}\frac{\partial}{\partial\eta}\left[(1-\eta^2)\frac{\partial\Psi_2}{\partial\eta}\right]$$

$$+ \frac{1}{\Psi_3}\left[\frac{1}{\zeta^2-1}+\frac{1}{1-\eta^2}\right]\frac{\partial^2\Psi_3}{\partial\phi^2} + k^2 d^2(\zeta^2-\eta^2) = 0$$

Multiplying by $\left[\frac{1}{\zeta^2-1}+\frac{1}{1-\eta^2}\right] = \frac{(\zeta^2-1)(1-\eta^2)}{(\zeta^2-\eta^2)}$,

$$\frac{(\zeta^2-1)(1-\eta^2)}{(\zeta^2-\eta^2)}\left[\frac{1}{\Psi_1}\frac{\partial}{\partial\zeta}\left[(\zeta^2-1)\frac{\partial\Psi_1}{\partial\zeta}\right] + \frac{1}{\Psi_2}\frac{\partial}{\partial\eta}\left[(1-\eta^2)\frac{\partial\Psi_2}{\partial\eta}\right]\right]$$

$$+ k^2 d^2(\zeta^2-1)(1-\eta^2) = -\frac{1}{\Psi_3}\frac{\partial^2\Psi_3}{\partial\phi^2} = m^2$$

$$\frac{\partial^2\Psi_3}{\partial\phi^2} + m^2\Psi_3 = 0$$

$$\frac{1}{\Psi_1}\frac{\partial}{\partial\zeta}\left[(\zeta^2-1)\frac{\partial\Psi_1}{\partial\zeta}\right] + \frac{1}{\Psi_2}\frac{\partial}{\partial\eta}\left[(1-\eta^2)\frac{\partial\Psi_2}{\partial\eta}\right]$$

$$- \frac{(\zeta^2-\eta^2)m^2}{(\zeta^2-1)(1-\eta^2)} + k^2 d^2(\zeta^2-\eta^2) = 0$$

or

$$\frac{1}{\Psi_1}\frac{\partial}{\partial\zeta}\left[(\zeta^2-1)\frac{\partial\Psi_1}{\partial\zeta}\right] + \left(k^2 d^2\zeta^2 - \frac{m^2}{\zeta^2-1}\right)$$

$$= -\frac{1}{\Psi_2}\frac{\partial}{\partial\eta}\left[(1-\eta^2)\frac{\partial\Psi_2}{\partial\eta}\right] + \left(k^2 d^2\eta^2 + \frac{m^2}{1-\eta^2}\right) = c$$

Hence

$$\frac{\partial}{\partial\zeta}\left[(\zeta^2-1)\frac{\partial\Psi_1}{\partial\zeta}\right] + \left(k^2 d^2\zeta^2 - \frac{m^2}{\zeta^2-1} - c\right)\Psi_1 = 0$$

$$\frac{\partial}{\partial\eta}\left[(1-\eta^2)\frac{\partial\Psi_2}{\partial\eta}\right] - \left(k^2 d^2\eta^2 + \frac{m^2}{1-\eta^2} - c\right)\Psi_2 = 0$$

Prob. 2.31 (a)

$$V(x,y,z) = \int_{m=1}^\infty \int_{n=1}^\infty \int_{p=1}^\infty A_{mnp}\sin mx \sin ny \sin pz$$

If we let $k_{mnp} = \dfrac{8}{\pi^3(m^2 + n^2 + p^2)}$,

$$A_{mnp} = k_{mnp}\int_0^\pi e^x \sin mx\,dx \int_0^\pi \sin ny\,dy \int_0^\pi \sin pz\,dz$$

$$= k_{mnp}\left(-\frac{1}{n}\cos ny\Big|_0^\pi\right)\left(-\frac{1}{p}\cos pz\Big|_0^\pi\right)\left(e^x \sin mx - \frac{m\cos mx}{m^2+1}\Big|_0^\pi\right)$$

$$= k_{mnp}\frac{2}{2n-1}\frac{2}{2p-1}\frac{m[1-(-1)^m e^\pi]}{m^2+1}$$

$$= \frac{32}{\pi^3[m^2+(2n-1)^2+(2p-1)^2]}\frac{m[1-(-1)^m e^\pi]}{(m^2-1)(2n-1)(2p-1)}$$

Hence

$$V(x,y,z) = \frac{32}{\pi^3}\int_{m=1}^\infty\int_{n=1}^\infty\int_{p=1}^\infty \frac{m[1-(-1)^m e^\pi]}{(m^2-1)(2n-1)(2p-1)}\cdot$$

$$\frac{\sin mx \sin(2n-1)y \sin(2p-1)z}{[m^2+(2n-1)^2+(2p-1)^2]}$$

(b)

$$A_{mnp} = k_{mnp}\int_0^\pi \sin^2 x \sin mx\,dx \int_0^\pi \sin ny\,dy \int_0^\pi \sin pz\,dz$$

$$= k_{mnp}\left(-\frac{1}{n}\cos ny\Big|_0^\pi\right)\left(-\frac{1}{p}\cos pz\Big|_0^\pi\right)\frac{1}{2}\int_0^\pi\left(\sin mx\,dx - \sin mx\cos 2x\,dx\right)$$

$$= \frac{2k_{mnp}}{(2n-1)(2p-1)}\begin{cases} 0, & m = \text{even} \\ -\frac{2}{m}+\frac{2}{2(m-2)}+\frac{2}{2(m+2)}, & m = \text{odd}\end{cases}$$

$$= \frac{16}{\pi^3[(2m-1)^3+(2m-1)^2+(2p-1)^2]}\left[-\frac{2}{m}+\frac{2}{2m-4}+\frac{2}{2m+4}\right]$$

Hence,

$$V(x,y,z) = -\frac{128}{\pi^3}\int_{m=1}^\infty\int_{n=1}^\infty\int_{p=1}^\infty \frac{1}{(2m-3)(4m^2-1)(2n-1)(2p-1)}\cdot$$

$$\frac{\sin(2m-1)x \sin(2n-1)y \sin(2p-1)z}{[(2m-1)^2+(2n-1)^2+(2p-1)^2]}$$

Prob. 2.32 First, solve the homogeneous equation

$$\nabla^2\Phi - \frac{\partial^2\Phi}{\partial^2 t} = 0$$

Let $\Phi(\rho, t) = R(\rho)T(t)$,

$$T'' + \lambda^2 T = 0, \qquad \rho^2 R'' + \rho R' + \lambda^2 \rho^2 R = 0$$

which have solutions

$$T_n(t) = a_n \cos \lambda t + b_n \sin \lambda t, \qquad R_n(\rho) = c_n J_0(\lambda \rho) + d_n Y_o(\lambda \rho)$$

Since $r = 0$ in included, $d_n = 0$. Let

$$\Phi(\rho, t) = \sum_{n=1}^{\infty} A(t) J_0(\lambda \rho), \; \lambda = \lambda_{on}/a$$

$$\frac{\partial \Phi}{\partial \rho} = -\sum_{n=1}^{\infty} \lambda A(t) J_1(\lambda \rho)$$

$$\frac{\partial^2 \Phi}{\partial \rho^2} = -\sum_{n=1}^{\infty} \lambda^2/2 A(t) J_o(\lambda \rho) + \sum_{n=1}^{\infty} \frac{\lambda}{\rho} A(t) J_1(\lambda \rho) - \sum_{n=1}^{\infty} \lambda^2/2 A(t) J_o(\lambda \rho)$$

Hence

$$\frac{\partial^2 \Phi}{\partial \rho^2} + \frac{1}{\rho} \frac{\partial \Phi}{\partial \rho} - \frac{\partial^2 \Phi}{\partial t^2} = -\sum_{n=1}^{\infty} \lambda^2 A(t) J_o(\lambda \rho)$$

$$+ \sum_{n=1}^{\infty} \frac{\lambda}{\rho} A(t) J_1(\lambda \rho) - \sum_{n=1}^{\infty} \frac{\lambda}{\rho} A(t) J_1(\lambda \rho) - \sum_{n=1}^{\infty} A''(t) J_o(\lambda \rho) = -\Phi_o \sin \omega t$$

or

$$\Phi_o \sin \omega t = \sum_{n=1}^{\infty} [A''(t) + \lambda_n^2 A(t)] J_o(\lambda \rho)$$

Multiply both sides by $\rho J(\lambda_{0m}\rho/a)$ and integrate over $0 < \rho < a$, we obtain

$$\sum_{n=1}^{\infty} \left[A''(t) + \frac{\lambda_{0n}^2}{a^2} A(t) \right] \int_0^a \rho J_0(\lambda_{0n}\rho/a) J_0(\lambda_{0m}\rho/a) d\rho = \Phi_o \sin \omega t \int_0^a J_0(\lambda_{0m}\rho/a) d\rho$$

$$\left[A''(t) + \frac{\lambda_{0n}^2}{a^2} A(t) \right] \frac{a^2}{2} J_1^2(\lambda_{on}) = \frac{a}{\lambda_{0n}} J_1(\lambda_{0n}) \Phi_o \sin \omega t$$

or

$$A''(t) + \frac{\lambda_{0n}^2}{a^2} A(t) = k \sin \omega t, \qquad k = \frac{2\Phi_o}{a \lambda_{0n} J_1(\lambda_{0n})}$$

This is an inhomogeneous, ordinary differential equation. Let $A = A_h + A_p$.

$$A_h''(t) + \frac{\lambda_{0n}^2}{a^2} A_h = 0$$

$$A_h = c_1 \cos(\lambda_{0n} t/a) + c_s \sin(\lambda_{0n} t/a)$$

Let

$$A_p = c_3 \sin \omega t + c_4 \cos \omega t$$

$$-\omega^2 c_3 \sin \omega t + \frac{\lambda_{0n}^2}{a^2} c_3 \sin \omega t - \omega^2 c_4 \cos \omega t + \frac{\lambda_{0n}^2}{a^2} c_4 \sin \omega t = k \sin \omega t$$

Equating components, $c_4 = 0$,

$$c_3 = \frac{k}{\frac{\lambda_{0n}^2}{a^2} - \omega^2}$$

Hence

$$A(t) = c_1 \cos(\lambda_{0n} t/a) + c_s \sin(\lambda_{0n} t/a) + \frac{k \sin \omega t}{\frac{\lambda_{0n}^2}{a^2} - \omega^2}$$

But

$$\Phi(\rho, 0) = 0 \quad \rightarrow \quad A(0) = 0 \quad \rightarrow \quad c_1 = 0$$

$$\Phi_t(\rho, 0) = 0 \quad \rightarrow \quad 0 = \frac{\lambda_{on} c_2}{a} + \frac{\omega k}{\lambda_{0n}^2/a^2 - \omega^2}$$

$$c_2 = -\frac{\omega k a \lambda_{0n}}{\lambda_{0n}^2/a^2 - \omega^2}$$

Thus,

$$A(t) = \frac{2a^2 \Phi_o}{\lambda_{on}^2 J_1(\lambda_{0n})} \frac{-\omega \sin(\lambda_{0n} t/a) + \frac{\lambda_{0n}}{a} \sin \omega t}{\lambda_{0n}^2 - a^2 \omega^2}$$

$$\Phi(\rho, t) = 2a^2 \Phi_o \sum_{n=1}^{\infty} \frac{J_0(\lambda_{0n} \rho/a)}{\lambda_{on}^2 J_1(\lambda_{0n})} \frac{\left(\frac{\lambda_{0n}}{a} \sin \omega t - \omega \sin(\lambda_{0n} t/a) \right)}{\lambda_{0n}^2 - a^2 \omega^2}$$

Prob. 2.33 We are to solve

$$\nabla^2 V = -\frac{\rho_v}{\epsilon} = -f(x, y) = -\frac{\rho_o x}{\epsilon a}$$

subject to

$$V(0, y) = 0 = V(a, y) = V(x, 0) = V(x, b)$$

Let

$$V(x, y) = \sum_{m=1}^{\infty} \sum_{n=1}^{\infty} A_{mn} \sin(m\pi x/a) \sin(n\pi y/b)$$

We substitute this into Poisson's equation

$$\frac{\partial^2 V}{\partial x^2} + \frac{\partial^2 V}{\partial y^2} = -f(x, y)$$

$$- \sum \sum A_{mn}(m\pi/a)^2 \sin(m\pi x/a)\sin(n\pi y/b)$$

$$- \sum \sum A_{mn}(n\pi/b)^2 \sin(m\pi x/a)\sin(n\pi y/b) = -f(x,y)$$

Multiplying both sides by $\sin(i\pi x/a)\sin(j\pi y/b)$ and integrating over $0 < x < a,\ 0 < y < b$, we get

$$A_{mn}\left[(m\pi/a)^2 + (n\pi/b)^2\right]\frac{a}{2}\frac{b}{2} = \int_0^b \int_0^a f(x,y)\sin(i\pi x/a)\sin(j\pi y/b)dxdy$$

$$A_{mn} = \frac{4}{ab}\frac{1}{\left[(m\pi/a)^2 + (n\pi/b)^2\right]} \cdot$$

$$\frac{\rho_o}{\epsilon a}\int_0^b \sin(n\pi y/b)dy \int_0^a x\sin(m\pi x/a)dx$$

But

$$\int_0^b \sin(n\pi y/b)dy = -\frac{b}{n\pi}(\cos n\pi - 1)$$

$$\int_0^a x\sin(m\pi x/a)dx = -\frac{a}{m\pi}a\cos m\pi$$

$$A_{mn} = \frac{\rho_o}{\epsilon a}\frac{4}{ab}\frac{b}{n\pi}\frac{a}{m\pi}\frac{a\cos m\pi(\cos n\pi - 1)}{\left[(m\pi/a)^2 + (n\pi/b)^2\right]}$$

Thus,

$$V(x,y) = \sum_{m=0}^\infty \sum_{n=0}^\infty \frac{4\rho_o \cos m\pi(\cos n\pi - 1)}{\pi^2 \epsilon mn\left[(m\pi/a)^2 + (n\pi/b)^2\right]} \cdot sin(m\pi x/a)\sin(n\pi y/b)$$

Prob. 2.34 From Section 2.7.2,

$$\mathbf{E}_g = -\nabla V_g = -\frac{\partial V_g}{\partial \rho}\mathbf{a}_\rho - \frac{\partial V_g}{\partial z}\mathbf{a}_z$$

Note that

$$\frac{d}{dx}J_0(x) = \frac{1}{2}[J_{-1}(x) - J_1(x)] = -J_1(x)$$

$$\frac{d}{d\rho}[J_0(\lambda_n\rho)] = -\lambda_n J_1(\lambda_n\rho)$$

$$\frac{d}{dz}\sinh[\lambda_n(b+c-z)] = -\lambda_n\cosh[\lambda_c(b+c-z)]$$

Hence

$$\mathbf{E}_g = \mathbf{a}_\rho \rho_v \sum_{n=1}^{\infty} \frac{2J_1(\lambda_n \rho)\lambda_n}{C_n K_n}[\cosh(\lambda_n b) - 1]\sinh[\lambda_n(b + c - z)]$$

$$+ \mathbf{a}_z \rho_v \sum_{n=1}^{\infty} \frac{2J_0(\lambda_n \rho)\lambda_n}{C_n K_n}[\cosh(\lambda_n b) - 1]\cosh[\lambda_n(b + c - z)]$$

Similarly,

$$\mathbf{E}_l = -\nabla V_l = -\frac{\partial V_l}{\partial \rho}\mathbf{a}_\rho - \frac{\partial V_l}{\partial z}\mathbf{a}_z$$

$$= \mathbf{a}_\rho \rho_v \sum_{n=1}^{\infty} \frac{2\lambda_n J_1(\lambda_n \rho)}{\epsilon_r C_n}\left[\frac{\sinh(\lambda_n z)}{K_n}T_n - \cosh(\lambda_n z) + 1\right]$$

$$- \mathbf{a}_\rho \rho_v \sum_{n=1}^{\infty} \frac{2\lambda_n J_1(\lambda_n \rho)}{\epsilon_r C_n\left[\frac{\cosh(\lambda_n z)}{K_n}T_n - \sinh(\lambda_n z)\right]}$$

where $T_n = \cosh(\lambda_n b)\cosh(\lambda_n c) + \epsilon_r \sinh(\lambda_n b)\sinh(\lambda_n c) - \cosh(\lambda_n c)$.

Prob. 2.35 Let

$$V_1 = \sum_{n=1}^{\infty} \sin\beta x\left[a_n \sinh\beta y + b_n \cosh\beta y\right], \quad c < y < b \tag{1}$$

$$V_2 = \sum_{n=1}^{\infty} c_n \sin\beta x \sinh\beta y, \quad 0 < y < c \tag{2}$$

where $\beta = n\pi/a$. At the interface, $y = c$, $V_1 = V_2$ and $\epsilon_1 \frac{\partial V_1}{\partial y} = \epsilon_2 \frac{\partial V_2}{\partial y}$.

$$V_1(y = c) = V_2(y = c) \quad \rightarrow \quad c_n = a_n + b_n \coth\beta c \tag{3}$$

$$\epsilon_1 \frac{\partial V_1}{\partial y}\bigg|_{y=c} = \epsilon_2 \frac{\partial V_2}{\partial y}\bigg|_{y=c} \quad \rightarrow$$
$$\epsilon_2 c_n \cosh\beta c = \epsilon_1 a_n \cosh\beta c + \epsilon_1 b_n \sinh\beta c \tag{4}$$

$$V_1(y = b) = V_o = \sum_{n=1}^{\infty} \sin\beta x\left[a_n \sinh\beta b + b_n \cosh\beta b\right]$$

Multiplying both sides by $\sin(m\pi x/a)$ and integrating over $(0, a)$ gives

$$a_n \sinh\beta b + b_n \cosh\beta b = \frac{4V_o}{n\pi} = K_o, \quad n = 1, 3, 5, \cdots \tag{5}$$

or

$$a_n = \frac{K_o}{\sinh \beta b} - b_n \coth \beta b \tag{6}$$

Substituting (3) and (6) into (4) gives

$$b_n = \frac{4V_o(\epsilon_2 - \epsilon_1)}{n\pi \sinh \beta b \left[\epsilon_1 \tanh \beta c - \epsilon_2 \coth \beta c + (\epsilon_2 - \epsilon_1) \coth \beta b\right]} \tag{7}$$

Substituting (7) into (6),

$$a_n = \frac{4V_o \left[\epsilon_1 \tanh \beta c - \epsilon_2 \coth \beta c\right]}{n\pi \sinh \beta b \left[\epsilon_1 \tanh \beta c - \epsilon_2 \coth \beta c + (\epsilon_2 - \epsilon_1) \coth \beta b\right]} \tag{8}$$

Substituting (7) and (8) into (3) gives

$$c_n = \frac{4V_o \left[\epsilon_1 \tanh \beta c - \epsilon_2 \coth \beta c + (\epsilon_2 - \epsilon_1) \coth \beta c\right]}{n\pi \sinh \beta b \left[\epsilon_1 \tanh \beta c - \epsilon_2 \coth \beta c + (\epsilon_2 - \epsilon_1) \coth \beta b\right]} \tag{8}$$

Check: If $\epsilon_2 = \epsilon_1 \quad \rightarrow \quad b_n = 0, \ c_n = a_n = \frac{4V_o}{n\pi \sinh \beta b}$.

Prob. 2.36 To solve the problem

$$\nabla^2 V_g = 0, \quad 0 \le x \le a, \ b \le y \le c \tag{1}$$

$$\nabla^2 V_\ell = -\frac{\rho}{\epsilon}, \quad 0 \le x \le a, \ 0 \le y \le b \tag{2}$$

Let

$$V_g = \sum_{n=1}^{\infty} A_n \sin \beta x \sinh \beta(c - y) \tag{3}$$

$$V_\ell = \sum_{n=1}^{\infty} \sin \beta x F_n(y) \tag{4}$$

where $\beta = n\pi/a$. Substituting (4) into (2),

$$\sum_{n=1}^{\infty} \sin \beta x G_n = -\rho/\epsilon \tag{5}$$

where

$$G_n = F_n'' - \beta^2 F_n \tag{6}$$

Multiplying both sides of (5) by $\sin(m\pi x/a)$ and integrating over $0 < x < a$ yields

$$G_n = -\frac{2\rho}{n\pi\epsilon}[1 - (-1)^n]$$

Once G_n is known, we solve the inhomogeneous ordinary differential equation (6) by letting

$$F_n = F_h + F_p$$

where F_h is the homogeneous solution and F_p is the particular solution. By letting $F_p = $ constant and substituting this in (6) readily gives

$$F_p = -\frac{G_n}{\rho^2} = \begin{cases} 0, & n = \text{even} \\ \frac{4\rho a^2}{n^3\pi^3\epsilon}, & n = \text{odd} \end{cases} \tag{7}$$

If we let $F_h = D_n \cosh\beta y + C_n \sinh\beta y$, then (4) becomes

$$V_\ell = \sum_{n=1}^{\infty} \sin\beta x\left[C_n \sinh\beta y + D_n \cosh\beta y + F_p\right] \tag{8}$$

We now impose the boundary condition to determine A_n, C_n, and D_n. At $y = 0$, $V_\ell = 0$, $\rightarrow$ $D_n = -F_p$ so that (8) becomes

$$V_\ell = \sum_{n=1}^{\infty} \sin\beta x\left[C_n \sinh\beta y - F_p(\cosh\beta y - 1)\right] \tag{9}$$

At $y = b$, $V_\ell = V_g$, $\rightarrow$

$$A_n \sinh\beta(c - b) = C_n \sinh\beta b - F_p(\cosh\beta b - 1) \tag{10}$$

Also,

$$\frac{\partial V_g}{\partial y}\bigg|_{y=b} = \epsilon_r \frac{\partial V_\ell}{\partial y}\bigg|_{y=b}$$

or

$$-A_n \cosh\beta(c - b) = \epsilon_r\left[C_n \cosh\beta b - F_p \sinh\beta b\right] \tag{11}$$

Solving (10) and (11) gives

$$C_n = F_p\left[\frac{\epsilon_r \sinh\beta b \sinh\beta(c - b) + \cosh\beta(c - b)[\cosh\beta b - 1]}{\epsilon_r \cosh\beta b \sinh\beta(c - b) + \sinh\beta b \cosh\beta(c - b)}\right] \tag{12}$$

$$A_n = C_n \frac{\sinh\beta b}{\sinh\beta(c - b)} - F_p \frac{(\cosh\beta b - 1)}{\sinh\beta(c - b)} \tag{13}$$

where F_p is given by (7). Substitution of (12) and (13) into (3) and (9) completes the solution.

Prob. 2.37 We may use the series expansion in Table 2.4, item (d). Alternatively, we may use the recurrence relations. Using $P_0^0(x) = 1.0$ and $P_1^0(x) = x$ as initial terms, higher terms such as P_2^0, P_3^0, P_4^0, etc. are generated with

$$P_n(x) = \frac{(2n-1)xP_{n-1}^m(x) - (n+m-1)P_{n-2}^m(x)}{n-m}$$

By using $P_1^1(x) = \sqrt{1-x^2}$ and $P_2^1(x) = 3x\sqrt{1-x^2}$ as initial terms P_3^1, P_4^1, P_5^1, etc. are generated with the above equation. For $m > 1$, we use

$$P_n^m(x) = \frac{2(m-1)xP_n^{m-1}(x)}{\sqrt{1-x^2}} - (n-m+2)(n+m-1)P_n^{m-2}$$

Thus with four initial terms, P_n^m can be generated using the recurrence relations for fixed n and m respectively. Typically, for $x = 0.5$,

$$P_1^0 = 0.5, \ P_2^1 = 1.29904, \ P_2^0 = -1.25, \ P_2^2 = 2.25$$
$$P_3^0 = -0.4375, \ P_3^1 = 0.32476, \ P_3^2 = 5.625, \ P_2^2 = 9.74279$$

Prob. 2.38 With the initial terms

$$j_0(x) = \frac{\sin x}{x}, \ j_1(x) = \frac{\sin x}{x^2} - \frac{\cos x}{x} = \frac{j_0(x) - \cos x}{x}$$

we use the recurrence relation

$$j_n(x) = \frac{(2n-1)}{x}j_{n-1}(x) - j_{n-2}(x)$$

to generate higher-order terms.

Prob. 2.39 For the Bessel function,

$$G(x,t) = \exp[\frac{x}{2}(t - \frac{1}{t})] = \sum_{n=-\infty}^{\infty} t^n J_n(x)$$

If $G(x+y,t) = G(x,t)G(y,t)$, then

$$\exp[\frac{x+y}{2}(t - \frac{1}{t})] = \exp[\frac{x}{2}(t - \frac{1}{t})]\exp[\frac{y}{2}(t - \frac{1}{t})]$$

The left-hand side (LHS) is given by

$$\text{LHS} = \sum_{n=-\infty}^{\infty} t^n J_n(x+y)$$

while the right-hand side (RHS) is given by

$$\text{RHS} = \Big[\sum_{m=-\infty}^{\infty} t^m J_m(x) \Big]\Big[\sum_{k=-\infty}^{\infty} t^k J_k(y) \Big]$$

i.e.

$$\sum_{n=-\infty}^{\infty} t^n J_n(x+y) = \sum_{k=-\infty}^{\infty} \sum_{m=-\infty}^{\infty} t^{m+k} J_m(x) J_k(y)$$

On the RHS, we let $m+k=n \;\rightarrow\; k=n-m$

$$\sum_{n=-\infty}^{\infty} t^n J_n(x+y) = \sum_{n=-\infty}^{\infty} \sum_{m=-\infty}^{\infty} t^n J_m(x) J_{n-m}(y)$$

Thus,

$$J_n(x+y) = \sum_{m=-\infty}^{\infty} J_m(x) J_{n-m}(y)$$

Prob. 2.40

$$\frac{1}{(1-2xt+t^2)^{1/2}} = \sum_{n=0}^{\infty} t^n P_n(x)$$

If $t=r_o/r,\; x=\cos\alpha$,

$$\frac{1}{R} = \frac{1}{r}\frac{1}{\left[1-\frac{2r_o}{r}\cos\alpha+\frac{r_o^2}{r^2}\right]} = \frac{1}{r}\sum_{n=0}^{\infty}(r_o/r)^2 P_n(\cos\alpha), \quad r>r_o$$

or

$$\frac{1}{R} = \frac{1}{r_o}\frac{1}{\left[1-\frac{2r}{r_o}\cos\alpha+\frac{r^2}{r_o^2}\right]} = \frac{1}{r_o}\sum_{n=0}^{\infty}(r/r_o)^2 P_n(\cos\alpha), \quad r<r_o$$

Prob. 2.41 (a) $\int T_0(x)dx = \int 1dx = x = T_1(x)$

(b) $\int T_1(x)dx = \int xdx = \frac{x^2}{2} = \frac{1}{4}[T_2(x)+1]$

(c) $\int T_n(x)dx = \int \cos(n\cos^{-1}x)dx$. Let $u=\cos^{-1}x,\; x=\cos u,\; dx=-\sin u\,du$

$$\int T_n(x)dx = -\int \cos(nu)\sin u\,du$$

But $\sin u\cos u = \frac{1}{2}[\sin(nu+u)+\sin(u-nu)]$.

$$T_n(x)dx = -\frac{1}{2}\int[\sin(nu+u)+\sin(u-nu)]du$$
$$= \frac{1}{2}\left[\frac{\cos u(n+1)}{n+1} - \frac{\cos u(n-1)}{n-1}\right]$$
$$= \frac{1}{2}\left[\frac{\cos[(n+1)\cos^{-1}(x)]}{n+1} - \frac{\cos[(n-1)\cos^{-1}(x)]}{n-1}\right]$$
$$= \frac{1}{2}\left[\frac{T_{n+1}}{n+1} - \frac{T_{n-1}}{n-1}\right]$$

48

Prob. 2.42 (a) Let $f(x) = \sum_{n=1}^{\infty} a_n H_n(x)$

$$\int_{\infty}^{\infty} f(x)H_m(x)e^{-x^2}dx = \sum_{n=1}^{\infty} a_n \int_{-\infty}^{\infty} e^{-x^2}H_n(x)H_m(x)dx = 2^n n!\sqrt{\pi}\delta_{mn}a_n$$

or

$$a_n = \frac{1}{2^n n!\sqrt{\pi}} \int_0^1 e^{-x^2}(1)H_n(x)dx$$

(b) Let $f(x) = \sum_{n=1}^{\infty} a_n L_n(x)$

$$\int_0^{\infty} f(x)L_m(x)e^{-x}dx = \sum_{n=1}^{\infty} a_n \int_0^{\infty} e^{-x}L_n(x)L_m(x)dx = a_n\delta_{mn}$$

$$a_n = \int_0^1 e^{-x}L_n(x)dx$$

$$a_0 = \int_0^1 e^{-x}dx = -e^{-x}\Big|_0^1 = 1 - 1/e$$

$$a_1 = \int_0^1 (1-x)e^{-x}dx = (1-1/e) + e^{-x}(x+1)\Big|_0^1 = 1/e$$

Prob. 2.43 Since

$$e^{jz} = e^{jz\cos\theta} = \sum_{n=0}^{\infty} j^n(2n+1)j_n(r)P_n(\cos\theta)$$

and

$$\mathbf{a}_x = \sin\theta\cos\phi\mathbf{a}_r + \cos\theta\cos\phi\mathbf{a}_\theta - \sin\phi\mathbf{a}_\phi$$

$$E_\theta^i = E_o e^{j\omega t}e^{-jkr\cos\theta}\cos\theta\cos\phi$$

$$= E_o e^{j\omega t}\sum_{n=0}^{\infty}(-j)^n(2n+1)j_n(kr)\cos\theta\cos\phi P_n(\cos\theta)$$

$$= E_o e^{j\omega t}\cos\phi\sum_{n=0}^{\infty}(-j)^n j_n(kr)\big[(n+1)P_{n+1}(x) + nP_{n-1}(x)\big], \quad x = \cos\theta$$

$$E_\phi^i = E_o e^{j\omega t} \sin\phi \sum_{n=0}^{\infty} (-j)^{n+1} j_n(kr) P_n(\cos\theta)$$

$$E_r^i = E_o e^{j\omega t} \frac{\cos\phi}{jkr} \sum_{n=0}^{\infty} (-j)^n (2n+1) j_n(kr) \frac{d}{d\theta} P_n(\cos\theta)$$

$$\mathbf{E}^i = E_o e^{j(\omega t - kz)} \mathbf{a}_x = E_r^i \mathbf{a}_r + E_\theta^i \mathbf{a}_\theta + E_\phi^i \mathbf{a}_\phi$$

We now compare this with eq. (2.158a), i.e.

$$\mathbf{E}^i = E_o e^{j\omega t} \sum_{n=1}^{\infty} (-j)^n \frac{2n+1}{n(n+1)} \left[M_n^{(1)}(k) + j N_n^{(1)}(k) \right]$$

$$M_n(k) = \frac{1}{\sin\theta} j_n(kr) P_n^1(\cos\theta) \cos\theta \mathbf{a}_\theta - j_n(kr) \frac{dP_n^1}{d\theta} \sin\phi \mathbf{a}_\phi$$

$$N_n(k) = \frac{n(n+1)}{kr} j_n(kr) P_n^1(\cos\theta) \cos\theta \mathbf{a}_r + \frac{1}{kr}\frac{d}{dr} j_n(kr) \frac{dP_n^1}{d\theta} \cos\phi \mathbf{a}_\theta$$

$$+ \frac{1}{kr\sin\theta}\frac{1}{kr}\frac{d}{dr} j_n(kr) \frac{dP_n^1}{d\theta} \sin\phi \mathbf{a}_\phi$$

But

$$\frac{d}{dr} j_n(kr) = \frac{1}{2}\left[j_{n-1}(kr) - j_{n+1}(kr) \right]$$

$$\frac{dP_n^1(\cos\theta)}{d\theta} = \frac{dP_n^1(\cos\theta)}{d\cos\theta} \cdot \frac{d}{d\theta}(\cos\theta) = (n+1)\cos\theta\left[\cos\theta P_n(\cos\theta) - P_{n+1}(\cos\theta) \right]$$

Comparing terms shows that Eq. (2.158a) is valid.

Prob. 2.44

$$\rho_n(x) = \frac{d}{dx} \ln[x h_n^{(2)}(x)] = \frac{[x h_n^{(2)}(x)]'}{x h_n^{(2)}(x)}$$

$$\sigma_n(x) = \frac{d}{dx} \ln[x j_n(x)] = \frac{[x j_n(x)]'}{x j_n(x)}$$

From Eq. (2.167a),

$$a_n = \frac{j_n(\alpha)}{h_n^{(2)}(\alpha)} \left[\frac{\frac{j_n(m\alpha)[\alpha j_n(\alpha)]'}{j_n(\alpha)} - [m\alpha j_n(m\alpha)]'}{\frac{j_n(m\alpha)[\alpha h_n^{(2)}(\alpha)]'}{h_n^{(2)}(\alpha)} - [m\alpha j_n(m\alpha)]'} \right]$$

$$= \frac{j_n(x)}{h_n^{(2)}(x)} \left[\frac{\frac{[\alpha j_n(\alpha)]'}{j_n(\alpha)} - \frac{[m\alpha j_n(m\alpha)]'}{m\alpha j_n(m\alpha)}}{\frac{[\alpha h_n^{(2)}(\alpha)]'}{h_n^{(2)}(\alpha)} - \frac{[m\alpha j_n(m\alpha)]'}{m\alpha j_n(m\alpha}} \right]$$

$$= \frac{j_n(x)}{h_n^{(2)}(x)} \left[\frac{\sigma_n(\alpha) - m\sigma_n(m\alpha)}{\rho_n(\alpha) - m\sigma(m\alpha)} \right]$$

Similarly, from Eq. (2.167b),

$$b_n = \frac{j_n(\alpha)}{h_n^{(2)}(\alpha)} \left[\frac{[m\alpha j_n(\alpha)]' - \frac{m^2 j_n(m\alpha)[\alpha j_n(\alpha)]'}{j_n(\alpha)}}{[m\alpha j_n(m\alpha)]' - \frac{m^2 j_n(m\alpha)[\alpha h_n^{(2)}(\alpha)]'}{h_n^{(2)}(\alpha)}} \right]$$

$$= \frac{j_n(x)}{h_n^{(2)}(x)} \left[\frac{\frac{[m\alpha j_n(m\alpha)]'}{m\alpha j_n(\alpha)} - \frac{[m\alpha j_n(\alpha)]'}{\alpha j_n(\alpha)}}{\frac{[m\alpha j_n(m\alpha)]'}{m\alpha j_n(m\alpha)} - \frac{m[\alpha h_n^{(2)}(\alpha)]'}{h_n^{(2)}(\alpha)}} \right]$$

$$= \frac{j_n(x)}{h_n^{(2)}(x)} \left[\frac{\sigma_n(m\alpha) - m\sigma_n(\alpha)}{\rho_n(m\alpha) - m\sigma(\alpha)} \right]$$

Prob. 2.45 Let $|E_x^i| = 1$

$$\mathbf{E}^t = E_o e^{j\omega t} \sum_{n=1}^{\infty} (-j)^n \frac{(2n+1)}{n(n+1)} \left[c_n \mathbf{M}_n^{(1)}(k_1) + j d_n \mathbf{N}_n^{(1)}(k_1) \right]$$

$$|e^{j\omega t}| = 1, \quad k_1 = \omega\sqrt{\mu_o \epsilon_o \mu_{r1} \epsilon_{r1}} = 2\omega/c = \frac{2 \times 2\pi \times 5 \times 10^9}{3 \times 10^8}$$

$$k_1 = 200\pi/3, \quad k = \omega/c = 100\pi/3, \quad m = k_1/k = 2$$

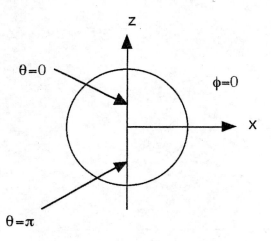

Since $\theta = 0$ or π,

$$E_x^t = E_r^t \cos\phi \sin\theta + E_\theta^t \cos\theta \cos\phi - E_\phi^t \sin\phi$$

$$= E_\theta^t = \begin{cases} E_\theta^t, & \theta = 0 \\ -E_\theta^t, & \theta = \pi \end{cases}$$

$$E_z^t = E_r^t \cos\theta - E_\theta^t \sin\theta = \begin{cases} E_r^t, & \theta = 0 \\ -E_r^t, & \theta = \pi \end{cases}$$

$$E_r^t = E_o e^{j\omega t} \sum_{n=1}^{\infty} (-j)^n \frac{(2n+1)}{n(n+1)} j d_n \frac{n(n+1)}{k_1 r} j_n(k_1 r) P_n^1(\cos\theta) \cos\phi$$

$$|E_r^t| = \left| \sum_{n=1}^{\infty} (-j)^{n+1} d_n \frac{2n+1}{k_1 r} j_n(k_1 r) P_n^1(\cos\theta) \right|$$

By using the program in Fig. 2.15 and varying $z = r$ between -a and a, the value of $|E_r^t|$ can be calculated.

Prob. 2.46 For the solution, see Fig. 4 of [17]. The plot is sketched below.

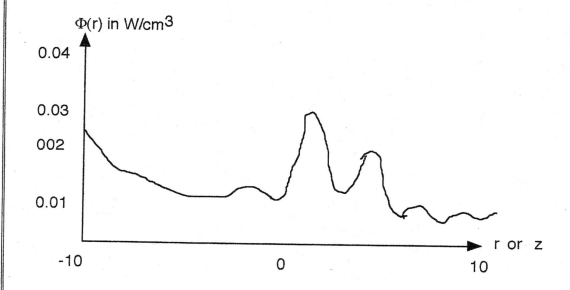

52

Prob. 2.47 The relative heating potential [18] is shown below

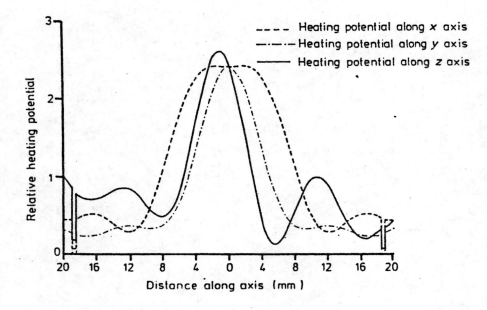

Prob. 2.48 The plot [19] is shown below - the homogeneous portion.

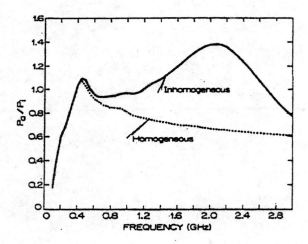

Prob. 2.49 The plot is shown above - the inhomogeneous portion [19].

CHAPTER 3

Prob. 3.1(a) Using Taylor series,

$$\Phi(x_o + \Delta x) \simeq \Phi(x_o) + \Delta x \Phi_x(x_o) \tag{1}$$
$$\Phi(x_o + 2\Delta x) \simeq \Phi(x_o) + 2\Delta x \Phi_x(x_o) \tag{2}$$

Multiplying (1) by 4 and subtracting (2) from it gives

$$4\Phi(x_o + \Delta x) - \Phi(x_o + 2\Delta x) = 3\Phi(x_o) + 2\Delta x \Phi_x(x_o)$$

or

$$\Phi_x(x_o) = \frac{4\Phi(x_o + \Delta x) - \Phi(x_o + 2\Delta x) - 3\Phi(x_o)}{2\Delta x}$$

which can be written as

$$\Phi_x = \frac{-\Phi_{i+2} + 4\Phi_{i+1} - 3\Phi_i}{2\Delta x}$$

(b)

$$\Phi(x_o - \Delta x) \simeq \Phi(x_o) - \Delta x \Phi_x(x_o) \tag{3}$$
$$\Phi(x_o - 2\Delta x) \simeq \Phi(x_o) - 2\Delta x \Phi_x(x_o) \tag{4}$$

Multiplying (3) by 4 and subtracting (4) leads to

$$\Phi_x = \frac{3\Phi_i - 4\Phi_{i-1} + \Phi_{i-2}}{2\Delta x}$$

(c) Multiplying (1) by 8 and subtracting 8 times (3) yields

$$8\Phi_{i+1} - 8\Phi_{i-1} = 16\Delta x \Phi_x \tag{5}$$

Subtracting (4) from (2),
$$\Phi_{i+2} - \Phi_{i-2} = 4\Delta x \Phi_x \tag{6}$$

Subtracting (6) from (5),

$$\Phi_x = \frac{-\Phi_{i+2} + 8\Phi_{i+1} - 8\Phi_{i-1} + \Phi_{i-2}}{12\Delta x}$$

(c) Using central difference,

$$\Phi_{i+1} - \Phi_{i-1} = 2\Delta x \Phi_x + \cdots \tag{7}$$
$$\Phi_{i+2} - \Phi_{i-2} = 4\Delta x \Phi_x + \cdots \tag{8}$$

Multiplying (7) by 8 and subtracting (4) from it,

$$-\Phi_{i+2} + 8\Phi_{i+1} - 8\Phi_{i-1} + \Phi_{i-2} = 12\Delta x \Phi_x$$

or

$$\Phi_x = \frac{-\Phi_{i+2} + 8\Phi_{i+1} - 8\Phi_{i-1} + \Phi_{i-2}}{12\Delta x}$$

Prob. 3.2 The exact solution is

$$\Phi(x,t) = \sin(\pi x)\exp(-\pi^2 t)$$

For the finite difference solution, $k = 1$, $r = 1/2$, $\Delta x = 1/4$, $\Delta t = rk(\Delta x)^2 = \frac{1}{32} = 0.03125$ With the explicit method,

$$\Phi(i, j+1) = \frac{1}{2}\big[\Phi(i+1,j) + \Phi(i-1,j)\big]$$

Applying this give the following results which are very close to the exact solution.

			Φ		
t/x	0	0.25	0.5	0.75	1.0
0	0	0.7071	1.0	0.7071	0
0.03125	0	0.5	0.7071	0.5	0
0.0625	0	0.3526	0.5	0.3536	0
0.09375	0	0.25	0.3526	0.25	0
0.125	0	0.1768	0.25	0.1768	0
0.15625	0	0.1250	0.1768	0.125	0
0.1875	0	0.0884	0.125	0.0884	0

Prob. 3.3

$$\Phi_{yy} = \frac{\Phi(i+1,j+1) - 2\Phi(i+1,j) + \Phi(i+1,j-1)}{2\Delta^2}$$
$$+ \frac{\Phi(i-1,j+1) - 2\Phi(i-1,j) + \Phi(i-1,j-1)}{2\Delta^2}$$
$$\Phi_{xx} = \frac{\Phi(i+1,j+1) - 2\Phi(i,j+1) + \Phi(i-1,j+1)}{2\Delta^2}$$
$$+ \frac{\Phi(i+1,j-1) - 2\Phi(i,j-1) + \Phi(i-1,j-1)}{2\Delta^2}$$

Substituting these in $\Phi_{xx} = a^2 \Phi_{yy}$ gives

$$(1 - a^2)\left[\Phi(i+1, j+1) + \Phi(i-1, j-1)\right] - \left[\Phi(i, j+1) + \Phi(i, j-1)\right.$$
$$\left. - a^2 \Phi(i+1, j) - a^2 \Phi(i-1, j)\right]$$
$$+ \Phi(i-1, j+1) + \Phi(i-1, j-1) - \Phi(i+1, j-1) - \Phi(i-1, j-1) = 0$$

Prob. 3.4 The exact solution is

$$\Phi = \frac{1}{6}x^3 + \frac{1}{2}x^2 + \frac{1}{3}x$$

so that $\Phi(0.5) = 0.3125$. For the finite difference solution,

$$\frac{\Phi(x + \Delta) - 2\Phi(x) + \Phi(x - \Delta)}{\Delta^2} = x + 1$$

which leads to

$$\Phi(x) = \frac{1}{2}\left[\Phi(x + \Delta) + \Phi(x - \Delta) - \Delta^2(x + 1)\right]$$

We apply this at $x = 0.25$, 0.5, 0.75 for 5 iterations as tabulated below.

No. of iterations	$\Phi(0)$	$\Phi(0.25)$	$\Phi(0.5)$	$\Phi(0.75)$	$\Phi(1.0)$
0	0	0	0	0	1
1	0	-0.03916	-0.0664	0.4121	1
2	0	-0.07226	0.1232	0.5069	1
3	0	0.02254	0.2178	0.5542	1
4	0	0.06984	0.2651	0.5779	1
5	0	0.09349	0.2888	0.5897	1

From the table, $\Phi(0.5) = 0.2888$ which is smaller than the exact value due to the fact that the number of iterations is not sufficiently large.

Prob. 3.5 From Table 3.1,

$$\Phi_{xx} = \frac{-\Phi(i+1, j) + 16\Phi(i+1, j) - 30\Phi(i, j) + 16\Phi(i-1, j) - \Phi(i-2, j)}{(\Delta x)^2}$$

Similarly,

$$\Phi_{yy} = \frac{-\Phi(i, j+2) + 16\Phi(i, j+1) - 30\Phi(i, j) + 16\Phi(i, j-1) - \Phi(i, j-1)}{(\Delta y)^2}$$

Let $\Delta x = \Delta y = \Delta$ and substitute the central formulas in $\Phi_{xx} + \Phi_{yy} = 0$ yields, after rearranging terms,

$$60\Phi(i,j) - 16\big[\Phi(i+1,j) + \Phi(i-1,j) + \Phi(i,j+1) + \Phi(i,j-1)\big]$$
$$+ \Phi(i+2,j) + \Phi(i-2,j) + \Phi(i,j+2) + \Phi(i,j-2) = 0$$

Prob. 3.6 From $\nabla^2 V = \frac{\partial^2 V}{\partial x^2} + \frac{\partial^2 V}{\partial x^2} = 0$,

$$\frac{V_1 - 2V_o + V_2}{(\Delta x)^2} + \frac{V_3 - 2V_o + V_4}{(\Delta x)^2} = 0$$

or

$$2V_o[1 + (\Delta x/\Delta y)^2] = V_1 + V_4 + (\Delta x/\Delta y)^2(V_3 + V_4)$$

Let $\alpha = (\Delta x/\Delta y)^2$. Then

$$V_o = \frac{V_1}{2(1+\alpha)} + \frac{V_2}{2(1+\alpha)} + \frac{V_3}{2(1+1/\alpha)} + \frac{V_4}{2(1+1/\alpha)}$$

(b)

$$V_{xx} = \frac{\frac{V_1 - V_o}{\Delta x_1} - \frac{V_o - V_2}{\Delta x_2}}{\frac{\Delta x_1 + \Delta x_2}{2}}$$
$$= \frac{2[-(\Delta x_1 + \Delta x_2)V_o + \Delta x_2 V_1 + \Delta x_1 V_2]}{\Delta x_1 \Delta x_2 (\Delta x_1 + \Delta x_2)}$$

Similarly,

$$V_{yy} = \frac{\frac{V_3 - V_o}{\Delta y_3} - \frac{V_o - V_4}{\Delta y_4}}{\frac{\Delta y_3 + \Delta y_4}{2}}$$
$$= \frac{2[-(\Delta y_3 + \Delta y_4)V_o + \Delta y_4 V_3 + \Delta y_3 V_4]}{\Delta y_3 \Delta y_4 (\Delta y_3 + \Delta y_4)}$$

Substituting these into $V_{xx} + V_{yy} = 0$ gives

$$\frac{-(\Delta x_1 + \Delta x_2)V_o + \Delta x_2 V_1 + \Delta x_1 V_2}{\Delta x_1 \Delta x_2 (\Delta x_1 + \Delta x_2)} + \frac{-(\Delta y_3 + \Delta y_4)V_o + \Delta y_4 V_3 + \Delta y_3 V_4]}{\Delta y_3 \Delta y_4 (\Delta y_3 + \Delta y_4)}$$

$$V_o\Big(\frac{1}{\Delta x_1 \Delta x_2} + \frac{1}{\Delta y_3 \Delta y_4}\Big) = \frac{V_1}{\Delta x_1 (\Delta x_1 + \Delta x_2)} + \frac{V_2}{\Delta x_2 (\Delta x_1 + \Delta x_2)}$$
$$+ \frac{V_3}{\Delta y_3 (\Delta y_3 + \Delta y_4)} + \frac{V_4}{\Delta y_4 (\Delta y_3 + \Delta y_4)}$$

or

$$V_o = \frac{V_1}{\left(1 + \frac{\Delta x_1}{\Delta x_2}\right)\left(1 + \frac{\Delta x_1 \Delta x_2}{\Delta y_3 \Delta y_4}\right)} + \frac{V_2}{\left(1 + \frac{\Delta x_2}{\Delta x_1}\right)\left(1 + \frac{\Delta x_1 \Delta x_2}{\Delta y_3 \Delta y_4}\right)}$$

$$+ \frac{V_3}{\left(1 + \frac{\Delta y_3}{\Delta y_4}\right)\left(1 + \frac{\Delta y_3 \Delta y_4}{\Delta x_1 \Delta x_2}\right)} + \frac{V_4}{\left(1 + \frac{\Delta y_4}{\Delta y_3}\right)\left(1 + \frac{\Delta y_3 \Delta y_4}{\Delta x_1 \Delta x_2}\right)}$$

(c) From eq. (3.26)

$$V_o = \frac{1}{4}(V_1 + V_2 + V_3 + V_4)$$

$$V_o = \frac{1}{4}(V_5 + V_6 + V_7 + V_8)$$

Adding gives

$$V_o = \frac{1}{8}(V_1 + V_2 + V_3 + V_4 + V_5 + V_6 + V_7 + V_8)$$

Alternatively, we can use Taylor series for V with two independent variables.

$$V(x + \Delta x, y + \Delta y) = V(x, y) + \Delta x \frac{\partial V}{\partial x} + \Delta y \frac{\partial V}{\partial y}$$

$$+ \frac{1}{2}\Delta x^2 \frac{\partial^2 V}{\partial x^2} + \frac{1}{2}\Delta y^2 \frac{\partial^2 V}{\partial y^2} + \Delta x \Delta y \frac{\partial^2 V}{\partial x \partial y} + \cdots$$

If $\Delta x = \Delta y = h$,

$$V(i+1, j+1) = V(i,j) + h\left(\frac{\partial V}{\partial x} + \frac{\partial V}{\partial x}\right) + \frac{h^2}{2}\left(\frac{\partial^2 V}{\partial x^2} + \frac{\partial^2 V}{\partial y^2}\right) + \cdots$$

$$V(i-1, j-1) = V(i,j) - h\left(\frac{\partial V}{\partial x} + \frac{\partial V}{\partial x}\right) + \frac{h^2}{2}\left(\frac{\partial^2 V}{\partial x^2} + \frac{\partial^2 V}{\partial y^2}\right) + \cdots$$

$$V(i-1, j+1) = V(i,j) - h\left(\frac{\partial V}{\partial x} - \frac{\partial V}{\partial x}\right) + \frac{h^2}{2}\nabla^2 V + \cdots$$

$$V(i+1, j-1) = V(i,j) + h\left(\frac{\partial V}{\partial x} - \frac{\partial V}{\partial x}\right) + \frac{h^2}{2}\nabla^2 V + \cdots$$

$$V(i+1, j) = V(i,j) + h\frac{\partial V}{\partial x} + \frac{h^2}{2}\frac{\partial^2 V}{\partial x^2} + \cdots$$

$$V(i-1, j) = V(i,j) - h\frac{\partial V}{\partial x} + \frac{h^2}{2}\frac{\partial^2 V}{\partial x^2} + \cdots$$

$$V(i, j-1) = V(i,j) - h\frac{\partial V}{\partial y} + \frac{h^2}{2}\frac{\partial^2 V}{\partial y^2} + \cdots$$

$$V(i, j+1) = V(i,j) + h\frac{\partial V}{\partial y} + \frac{h^2}{2}\frac{\partial^2 V}{\partial y^2} + \cdots$$

Adding all this gives

$$V_1 + V_2 + V_3 + V_4 + V_5 + V_6 + V_7 + V_8 = 8V_o + 3h^2 \nabla^2 V$$

If $\nabla^2 V = 0$,

$$V_o = \frac{1}{8}(V_1 + V_2 + V_3 + V_4 + V_5 + V_6 + V_7 + V_8)$$

Prob. 3.7

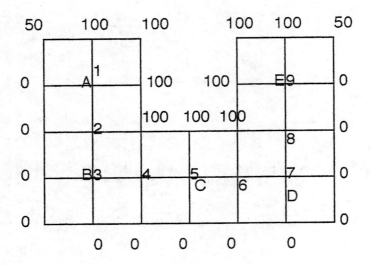

Applying

$$V(i,j) = \frac{1}{4}[V(i+1,j) + V(i-1,j) + V(i,j+1) + V(i,j-1)]$$

to all the free nodes gives

$$
\begin{bmatrix}
-4 & 1 & 0 & 0 & 0 & 0 & 0 & 0 & 0 \\
1 & -4 & 1 & 0 & 0 & 0 & 0 & 0 & 0 \\
0 & 1 & -4 & 1 & 0 & 0 & 0 & 0 & 0 \\
0 & 0 & 1 & -4 & 1 & 0 & 0 & 0 & 0 \\
0 & 0 & 0 & 1 & -4 & 1 & 0 & 0 & 0 \\
0 & 0 & 0 & 0 & 1 & -4 & 1 & 0 & 0 \\
0 & 0 & 0 & 0 & 0 & 1 & -4 & 1 & 0 \\
0 & 0 & 0 & 0 & 0 & 0 & 1 & -4 & 1 \\
0 & 0 & 0 & 0 & 0 & 0 & 0 & 1 & -4
\end{bmatrix}
\begin{bmatrix}
V_1 \\ V_2 \\ V_3 \\ V_4 \\ V_5 \\ V_6 \\ V_7 \\ V_8 \\ V_9
\end{bmatrix}
=
\begin{bmatrix}
-200 \\ -100 \\ 0 \\ -100 \\ -100 \\ -100 \\ 0 \\ -100 \\ -200
\end{bmatrix}
$$

or $AV = B$.

$$V = A^{-1}B = [61.46, 45.86, 21.96, 41.99, 45.99, 41.99, 21.96, 45.86, 61.46]^T$$

Thus,

$$V_A = 61.46, \ V_B = 21.96, \ V_C = 45.99, \ V_D = 21.96, \ V_E = 61.46$$

Prob. 3.8 Equation (3.5) applies to this problem, i.e.

$$\Phi(i,j) = \frac{1}{4}\left[\Phi(i+1,j+\Phi(i-1,j)+\Phi(i,j+1)+\Phi(i,j-1)+h^2 50\right] \quad (1)$$

At $y=0$,

$$\Phi_y = \frac{\Phi(i,1)-\Phi(i,-1)}{2h} = 40$$

or

$$\Phi(i,-1) = \Phi(i,1) - 80h \quad (2)$$

At $y=1$,

$$\Phi_y = \frac{\Phi(i,2)-\Phi(i,0)}{2h} = -20$$

or

$$\Phi(i,2) = \Phi(i,0) - 40h \quad (3)$$

Using (1) to (3) and other conditions, a program can be developed.

(a) For $h = 0.25$, the solution converges after 16 iterations and

$$\Phi(0.25, 0.25) = 8.8973$$

(b) For $h = 0.05$, the solution converges after 68 iterations and

$$\Phi(0.25, 0.25) = 8.649$$

For $h = 0.1$, the solution converges after 38 iterations and

$$\Phi(0.25, 0.25) = 8.6229$$

For $h = 0.2$, the solution converges after 20 iterations and

$$\Phi(0.25, 0.25) = 8.6698$$

Prob. 3.9 For the exact solution, let

$$V = V_1 + V_2$$

Using separation of variables and series expansion techniques, we obtain

$$V_1 = \frac{16V_o}{\pi^2}\sum_{m=1}^{\infty}\sum_{n=1}^{\infty}\frac{\sin j\pi x \sin k\pi y \sinh z\sqrt{j^2+k^2}}{jk \sinh \pi\sqrt{j^2+k^2}}, \quad j = 2m+1, k = 2n+1$$

$$V_2 = \frac{144}{\pi^3}\sum_{m=1}^{\infty}\sum_{n=1}^{\infty}\sum_{p=1}^{\infty}\frac{(-1)^{m+n}\left[\frac{2[(-1)^p-1]}{p^3\pi^3}-\frac{(-1)^p}{p\pi}\right]}{mn(m^2+n^2+p^2)}\cdot \sin m\pi x \sin n\pi y \sin p\pi z$$

For the finite difference solution with $\Delta x = \Delta y = \Delta z = h$,

$$\nabla^2 V = V_{xx} + V_{yy} + V_{zz} = g(x,y,z)$$

leads to

$$V(i,j,k) = \frac{1}{6}\Big[V(i+1,j,k) + V(i-1,j,k) + V(i,j+1,k) + V(i,j-1,k)$$
$$+ V(i,j,k+1) + V(i,j,k-1) - h^2 g(i,j,k)\Big]$$

Using this along with the boundary conditions on $x = 0, 1$, $y = 0, 1$, and $z = 0, 1$, a program can be developed. For the exact solution,

$$V(0.5, 0.5, 0.5) = 16.894$$

For the finite difference solution with $N_x = N_y = N_z = 4$,

$$V(0.5, 0.5, 0.5) = 16.667$$

Prob. 3.10 (a)
$$\Phi_i^{n+1} = \Phi_i^{n-1} + r(\Phi_{i+1}^n - 2\Phi_i^n + \Phi_{i-1}^n)$$

where $r = 2\Delta t/[k(\Delta x)^2]$. Let $\Phi_i^n = A^n e^{jkix}$.

$$A^{n+1}e^{jkix} = A^{n-1}e^{jkix} + rA^n e^{jkix}(e^{jkx} + e^{-jkx} - 2)$$

or

$$A^{n+1} = A^{n-1} + 2rA^n(\cos kx - 1)$$

i.e. $g^2 = 1 - 4pg$, $p = -\frac{r}{2}(\cos kx - 1) = r\sin^2 kx/2$

$$g^2 + 4pg - 1 = 0$$

$$g = \frac{-2p \pm \sqrt{16p^2 + 4}}{2} = -p \pm \sqrt{4p^2 + 1}$$

$$= -2r\sin^2(kx/2) \pm \sqrt{4r^2 \sin^2 kx/2 + 1}$$

$|g| > 1$, i.e. the scheme is unstable for all r.

(b)
$$\Phi_i^{n+1} = \Phi_i^{n-1} + r(\Phi_{i+1}^n + \Phi_{i-1}^n - \Phi_i^{n+1} + \Phi_i^{n-1})$$

$$A^{n+1} = A^{n-1} + rA^n[e^{jkx} + e^{-jkx}] - r[A^{n+1} + A^{n-1}]$$

$$= A^{n-1}\frac{(1-r)}{(1+r)} + \frac{2r}{1+r}A^n \cos kx$$

or

$$g^2 - \frac{2r}{1+r}g\cos kx - \frac{(1-r)}{(1+r)} = 0$$

From this, we obtain

$$g = \frac{1}{1+r}\left[r\cos kx \pm \sqrt{1 - 2r^2\sin^2 kx/2}\right]$$

$|g| \le 1$ for all values of r; i.e. the scheme is stable.

Prob. 3.11 (a) Let $\Phi_i^n = A^n e^{jki\Delta}$

$$A^{n+1}e^{jki\Delta} = A^n e^{jki\Delta} - rA^n[e^{jk\Delta} - e^{-jk\Delta}]e^{jki\Delta}$$

$$g = \frac{A^{n+1}}{A^n} = 1 - 2r\sin k\Delta = 1 - j\alpha$$

where $\alpha = 2r\sin k\Delta$.

$$|g|^2 = gg^* = 1 + \alpha^2 = 1 + 4r^2\sin^2 k\Delta \ge 1$$

showing that the algorithm is unstable.

(b) Similarly,

$$A^{n+1} = A^n[\cos k\Delta - 2jr\sin k\Delta]$$

so that $g = \cos k\Delta - j2r\sin k\Delta$

$$|g|^2 = gg^* = \cos^2 k\Delta + 4r\sin^2 k\Delta$$
$$= 1 - (1 - 4r^2)\sin^2 k\Delta$$

$|g|^2 \le 1 \quad \rightarrow \quad r \le 1/2$ or $\Delta \le \frac{\Delta x}{u}$.

Prob. 3.12 Let

$$U_{i,l}^n = A^n e^{jk_x ix}e^{jk_y ly}$$

(i)

$$A^{n+1}e^{jk_x ix}e^{jk_y ly} = A^n e^{jk_x ix}e^{jk_y ly} + r\big[A^n e^{jk_x(i-1)x}e^{jk_y ly} - 2A^n e^{jk_x ix}e^{jk_y ly}$$
$$+ A^n e^{jk_x(i+1)x}e^{jk_y ly} + A^n e^{jk_x ix}e^{jk_y(l-1)y} - 2A^n e^{jk_x ix}e^{jk_y ly}$$
$$+ A^n e^{jk_x ix}e^{jk_y(l+1)y}\big]$$

Dividing by $A^n e^{jk_x ix}e^{jk_y ly}$ gives

$$\frac{A^{n+1}}{A^n} = 1 + r\big[-4 + 2\cos k_x x + 2\cos k_y y\big]$$

Since $\frac{A^{n+1}}{A^n} = g$ and $|g| \le 1$,

$$\left| 12r\left(2 + \cos k_x x + \cos k_y y\right) \right| \le 1$$

$$1 - 2r(2 + 1 + 1) \le -1 \qquad \text{or} \qquad r \le 1/4$$

(ii) Similarly,

$$A^{n+1} = \left(1 + 2r\cos k_x x - 2r\right)\left(1 + 2r\cos k_y y - 2r\right)A^n$$

Hence

$$g = (1 + \lambda_1 r)(1 + \lambda_2 r)$$

where $\lambda_1 = 2(\cos k_x x - 1)$, $\lambda_2 = 2(\cos k_y y - 1)$.

$$|g| \le 1 \qquad \rightarrow \qquad -1 \le g \le 1$$

Since the maximum value of λ_1 or λ_2 is 0, the upper limit of $g \le 1$ is satisfied. For the lower limit,

$$-1 \le g = (1 + \lambda_1 r)(1 + \lambda_2 r)$$

$$\lambda_1 \lambda_2 r^2 + (\lambda_1 + \lambda_2)r + 2 \ge 0$$

The minimum value of λ_1 or λ_2 is -4. When $\lambda_1 = \lambda_2 = -4$, r is complex. Since r must be real, λ_1 and λ_2 cannot be minimum at the same time. If either λ_1 or λ_2 is minimum and the other is maximum, say $\lambda_1 = -4$, $\lambda_2 = 0$, then

$$-4r + 2 \ge 0 \qquad \rightarrow \qquad r = 1/2$$

Prob. 3.13 (a)

$$\nabla \cdot \mathbf{E} = 0$$
$$\nabla \cdot \mathbf{H} = 0$$
$$\nabla \times \mathbf{E} = -\mu \frac{\partial \mathbf{H}}{\partial t}$$
$$\nabla \times \mathbf{H} = \sigma \mathbf{E} + \epsilon \frac{\partial \mathbf{E}}{\partial t} + \mathbf{J}_s \simeq \sigma \mathbf{E} + \mathbf{J}_s$$

$$\nabla \times \nabla \times \mathbf{E} = -\mu \frac{\partial}{\partial t}(\nabla \times \mathbf{H}) = -\mu\sigma \frac{\partial \mathbf{E}}{\partial t} - \mu \frac{\partial \mathbf{J}_s}{\partial t}$$

$$\nabla(\nabla \cdot \mathbf{E}) - \nabla^2 \mathbf{E} = -\mu\sigma \frac{\partial \mathbf{E}}{\partial t} - \mu \frac{\partial \mathbf{J}_s}{\partial t}$$

$$\frac{\partial^2 \mathbf{E}}{\partial x^2} + \frac{\partial^2 \mathbf{E}}{\partial y^2} - \mu\sigma \frac{\partial \mathbf{E}}{\partial t} = \mu \frac{\partial \mathbf{J}_s}{\partial t}$$

From this,

$$\frac{\partial^2 E_z}{\partial x^2} + \frac{\partial^2 E_z}{\partial y^2} - \mu\sigma\frac{\partial E_z}{\partial t} = \mu\frac{\partial J_{zs}}{\partial t}$$

(b) If $\mathbf{J}_s = 0$ and $\Delta x = \Delta y = \Delta$, we obtain the following formulas.

(i) Euler:

$$\frac{E_{i,j}^{n+1} - E_{i,j}^n}{\Delta t} = \frac{E_{i+1,j}^n - 2E_{i,j}^n + E_{i-1,j}^n}{\mu\sigma\Delta^2} + \frac{E_{i,j+1}^n - 2E_{i,j}^n + E_{i,j-1}^n}{\mu\sigma\Delta^2}$$

$$= \frac{E_{i+1,j}^n + E_{i-1,j}^n + E_{i,j+1}^n + E_{i,j-1}^n - 4E_{i,j}^n}{\mu\sigma\Delta^2}$$

or

$$E_{i,j}^{n+1} = (1 - 4r)E_{i,j}^n + r\sum E_{i,j}^n, \quad r = \frac{\Delta}{\mu\sigma\Delta^2}$$

(ii) Leapfrog:

$$\frac{E_{i,j}^{n+1} - E_{i,j}^{n-1}}{2\Delta t} = \frac{E_{i+1,j}^n - 2E_{i,j}^n + E_{i-1,j}^n}{\mu\sigma\Delta^2} + \frac{E_{i,j+1}^n - 2E_{i,j}^n + E_{i,j-1}^n}{\mu\sigma\Delta^2}$$

$$= \frac{\sum E_{i,j}^n - 4E_{i,j}^n}{\mu\sigma\Delta^2}$$

or

$$E_{i,j}^{n+1} = E_{i,j}^{n-1} + 2r\left[\sum E_{i,j}^n - 4E_{i,j}^n\right]$$

(iii) Dufort-Frankel:

$$\frac{E_{i,j}^{n+1} - E_{i,j}^{n-1}}{2\Delta t} = \frac{E_{i+1,j}^n - E_{i,j}^{n-1} - E_{i,j}^{n+1} + E_{i-1,j}^n}{\mu\sigma\Delta^2} + \frac{E_{i,j+1}^n - E_{i,j}^n - E_{i,j}^n + E_{i-1,j}^{n+1}}{\mu\sigma\Delta^2}$$

$$= \frac{\sum E_{i,j}^n - 2E_{i,j}^n - 2E_{i,j}^{n-1}}{\mu\sigma\Delta^2}$$

$$(1 + 4r)E_{i,j}^{n+1} = 2rE_{i,j}^n - 4rE_{i,j}^{n-1} + E_{i,j}^{n-1}$$

or

$$E_{i,j}^{n+1} = \frac{1 - 4r}{1 + 4r}E_{i,j}^{n-1} + \frac{2r}{1 + 4r}\sum E_{i,j}^n$$

(c) Let $E_{i,j}^n = A_n \cos(k_x i\Delta)\cos(k_y j\Delta)$

(i) Euler:

$$A^{n+1} = (1 - 4r + r\lambda)A^n$$

where $\lambda = 2(\cos k_x \Delta + \cos k_y \Delta)$

$$g_E = \frac{A^{n+1}}{A^n} = 1 - 4r + r\lambda$$

$$|g_E| \leq 1 \quad \rightarrow \quad -1 \leq g_E \leq 1$$

Since $r > 0$ and $\lambda \leq 4$, $g_E \leq 1$ always. The limit $-1 \leq g_E$ requires that

$$1 + r(\lambda - 4) = g \geq -1 \quad \rightarrow \quad 2 \geq r(4 - \lambda)$$

or

$$r \leq \min\left(\frac{2}{4 - \lambda}\right)$$

Since the minimum value of λ is -4, $r \leq 4$ for stability.

(ii) Leapfrog: Let

$$g_L = \frac{A^{n+1}}{A^n} = \frac{A^n}{A^{n-1}}$$

$$g_L^2 - 2r(\lambda - 4)g_L - 1 = 0$$

$$g_L^+ = r(\lambda - 4) + \sqrt{r^2(\lambda - 4)^2 + 1}$$
$$g_L^- = r(\lambda - 4) - \sqrt{r^2(\lambda - 4)^2 + 1}$$

Since $g_L^+ + g_L^- = 2r(\lambda - 4)$, $g_L^+ g_L^- = -1$, it is impossible to satisfy $|g_L| \leq 1$ for all k_x and k_y. Hence the Leapfrog scheme is always *unstable*.

(iii) Dufort-Frankel:

$$g_{DF} = \frac{A^{n+1}}{A^n} = \frac{A^n}{A^{n-1}}$$

$$g_{DF}^+ = \frac{r\lambda + \sqrt{r^2(\lambda - 4)(\lambda + 4) + 1}}{1 + 4r}$$

$$g_{DF}^- = \frac{r\lambda - \sqrt{r^2(\lambda - 4)(\lambda + 4) + 1}}{1 + 4r}$$

Both $|g_{DF}^+|$ and $|g_{DF}^-|$ are bounded by 1 for all k_x and k_y so that the scheme is *unconditionally stable*.

Prob. 3.14 Substituting $E_x^n(k) = A^n e^{j\beta k\delta}$, $H_x^n(k) = \frac{A^n}{\eta} e^{j\beta k\delta}$ into

$$H_y^{n+1/2}(k + 1/2) = H_y^{n-1/2}(k + 1/2) + \frac{\delta t}{\mu\delta}[E_x^n(k + 1) - E_x^n(k + 1)]$$

gives

$$A^{n+1/2} = A^{n-1/2} + \eta R_b A^n [e^{-j\beta\delta/2} - e^{j\beta\delta/2}]$$

where $R_b = \frac{\delta t}{\mu\delta}$. Let $g = A^{n+1/2}/A^n = A^n/A^{n-1/2}$,

$$g^2 = 1 - j2g\eta R_b \sin\frac{1}{2}\beta\delta$$

But

$$\eta R_b = \sqrt{\frac{\mu}{\epsilon}}\frac{\delta t}{\mu\delta} = \frac{u\delta t}{\delta} = r$$

$$g^2 + 2pg - 1 = 0, \qquad p = jr\sin\beta\delta/2$$

$$g_1 = -p - \sqrt{p^2 + 1}, \quad g_2 = -p + \sqrt{P^2 + 1}$$

If $|g_i| \leq 1$, $i = 1, 2$, p must lie between $-j$ and j, i.e.

$$-j \leq jr\sin\beta\delta/2 \leq j$$

Prob. 3.15

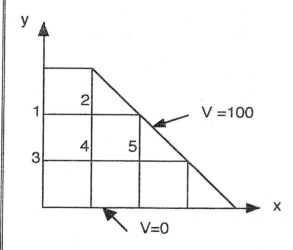

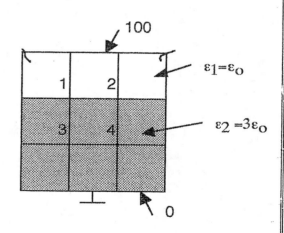

Applying the difference method,

$$V_1 = \frac{V_3}{4} + \frac{V_2}{2} + 25$$

$$V_2 = \frac{1}{4}(V_1 + V_4) + 50$$

$$V_3 = \frac{1}{4}(V_1 + V_4)$$

$$V_4 = \frac{1}{4}(V_2 + V_3 + V_5)$$

$$V_5 = \frac{V_4}{4} + 50$$

Applying this iteratively, we obtain the results below.

Iterations	0	1	2	3	4	5 $\cdots$	100
V_1	0	25.0	54.68	64.16	70.97	73.79	75.68
V_2	0	56.25	67.58	74.96	78.23	79.54	80.41
V_3	0	6.25	21.48	33.91	38.72	40.63	41.89
V_4	0	15.63	35.74	41.96	44.86	45.31	45.95
V_5	0	53.91	58.74	60.49	61.09	61.37	61.49

(b) On the interface,

$$\frac{\epsilon_1}{2(\epsilon_1 + \epsilon_2)} = \frac{1}{8}, \qquad \frac{\epsilon_2}{2(\epsilon_1 + \epsilon_2)} = \frac{3}{8}$$

$$V_1 = \frac{V_2}{4} + \frac{3V_3}{8} + 12.5$$

$$V_2 = 12.5 + \frac{3V_4}{8} + \frac{V_1}{4}$$

$$V_3 = \frac{1}{4}(V_1 + V_4)$$

$$V_4 = \frac{1}{4}(V_2 + V_3)$$

Applying this iteratively, we obtain

Iterations	0	1	2	3	4	5 $\cdots$	100
V_1	0	12.5	17.57	19.25	19.77	19.93	20
V_2	0	15.62	18.65	19.58	19.87	19.96	20
V_3	0	3.125	5.566	6.33	6.56	6.634	6.667
V_4	0	4.688	6.055	6.477	6.608	6.649	6.667

Prob. 3.16 By symmetry, $V_1 = V_5$, $V_4 = V_2$. Hence

$$2V_1 = 50 + V_2 \tag{1}$$
$$4V_2 = 100 + V_1 + V_3 \tag{2}$$
$$2V_3 = 100 + V_2 \tag{3}$$

Solving these simultaneously leads to

$$V_1 = 54.17 = V_5, \quad V_2 = 58.33 = V_4, \quad V_3 = 79.16$$

Prob. 3.18 The finite difference solution is obtained by following the same steps as in Example 3.6. We obtain $Z_o = 43 \ \Omega$.

Prob. 3.19 Due to double symmetry of the solution region, we consider only a quarter section and apply the appropriate finite difference formulas at the lines of symmetry. Let $n_x = A/\Delta, \ \Delta = \Delta x = \Delta y$.

(a)

$$n_x = 30 \quad \rightarrow \quad Z_o = 60.36 \ \Omega$$
$$n_x = 45 \quad \rightarrow \quad Z_o = 60.46 \ \Omega$$
$$n_x = 60 \quad \rightarrow \quad Z_o = 60.51 \ \Omega$$

(The exact solution gives $Z_o = 60.61 \ \Omega$.)

(b)

$$n_x = 18 \quad \rightarrow \quad Z_o = 49.87 \ \Omega$$
$$n_x = 36 \quad \rightarrow \quad Z_o = 50.28 \ \Omega$$
$$n_x = 72 \quad \rightarrow \quad Z_o = 50.44 \ \Omega$$

(The exact solution gives $Z_o = 50 \ \Omega$.)

Prob. 3.21 (a) exact: $k_c = 4.4429$ for $n_x = n_y$. The FDM gives

n_x	k_c
2	4.0
4	4.3296
6	4.3928
8	4.414
10	4.425

(b) exact: $k_c = 3.5124$ for $n_x = 2n_y$. The FDM gives

n_x	k_c
2	3.2163
4	3.436
6	3.478
8	3.493
10	3.5

Prob. 3.22 The results remain almost exactly the same as in Example 3.9.

Prob. 3.23(a) When $E_z = 0$ and $\frac{\partial}{\partial z} = 0$, $H_x = 0 = H_y$ and Maxwell's equations become

$$-\mu\frac{\partial H_z}{\partial t} = \frac{\partial E_y}{\partial x} - \frac{\partial E_x}{\partial y}$$

$$\frac{\partial H_z}{\partial y} = \epsilon\frac{\partial E_x}{\partial t}$$

$$-\frac{\partial H_z}{\partial t} = \epsilon\frac{\partial E_y}{\partial t}$$

Hence Yee algorithm becomes

$$H_z^{n+1/2}(i+1/2, j+1/2) = H_z^{n-1/2}(i+1/2, j+1/2) + \alpha\big[E_x^n(i+1/2, j+1)$$
$$- E_x^n(i+1/2, j)\big] + \alpha\big[E_y^n(i, j+1/2) - E_y^n(i+1, j+1/2)\big]$$
$$E_y^{n+1}(i+1/2, j) = \gamma E_y^n(i, j+1/2) + \beta\big[H_z^{n+1/2}(i-1/2, j+1/2)$$
$$- H_z^{n+1/2}(i+1/2, j+1/2)\big]$$
$$E_x^{n+1}(i+1/2, j) = \gamma E_x^n(i+1/2, j) + \beta\big[H_z^{n+1/2}(i+1/2, j+1/2)$$
$$- H_z^{n+1/2}(i+1/2, j-1/2)\big]$$

(a) When $H_z = 0$ and $\frac{\partial}{\partial z} = 0$, $E_x = 0 = E_y$ and Maxwell's equations become

$$\epsilon\frac{\partial E_z}{\partial t} = \frac{\partial H_y}{\partial x} - \frac{\partial H_x}{\partial y}$$

$$\mu\frac{\partial H_x}{\partial t} = -\frac{\partial E_z}{\partial t}$$

$$\mu\frac{\partial H_y}{\partial t} = \frac{\partial E_z}{\partial x}$$

Hence Yee's algorithm becomes

$$H_x^{n+1/2}(i, j+1/2) = H_x^{n-1/2}(i, j+1/2) + \alpha\big[E_z^n(i, j) - E_z^n(i, j+1)\big]$$
$$H_y^{n+1/2}(i+1/2, j) = H_y^{n+1/2}(i+1/2, j) + \alpha\big[E_z^n(i+1, j) - E_z^n(i, j)\big]$$
$$E_x^{n+1}(i, j) = \gamma E_x^n(i, j) + \beta\big[H_y^{n+1/2}(i+1/2, j) - H_y^{n+1/2}(i-1/2, j)\big]$$
$$+ \beta\big[H_x^{n+1/2}(i, j-1/2) - H_x^{n+1/2}(i, j+1/2)\big]$$

Prob. 3.24 Typically, E_z of the TM wave for $j = 30$ and time cycle $n = 95$ is plotted below [80, p. 37].

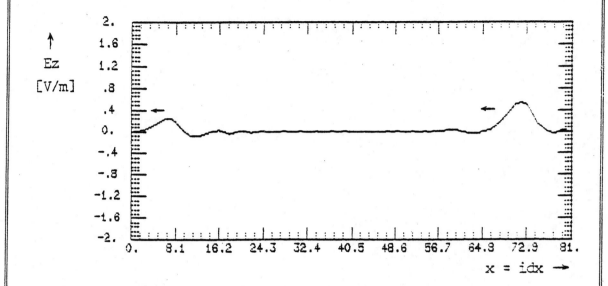

Prob. 3.25 Typically, E_z of the TM wave for $j = 30$ and time cycle $n = 95$ is plotted below [80, p. 90].

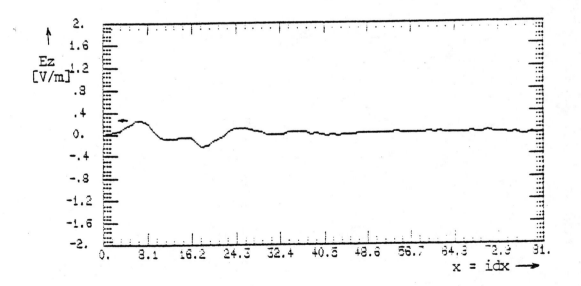

Prob. 3.26 $\nabla^2 A_z = -\mu J_z$ may be written as

$$\frac{\partial^2 A_z}{\partial \rho^2} + \frac{1}{\rho}\frac{\partial A_z}{\partial \rho} + \frac{1}{\rho^2}\frac{\partial^2 A_z}{\partial \phi^2} = -\mu J_z$$

The finite difference equivalent is

$$\frac{A_z(\rho+\Delta\rho,\phi) - 2A_z(\rho,\phi) + A_z(\rho-\Delta\rho,\phi)}{(\Delta\rho)^2} + \frac{1}{\rho}\frac{A_z(\rho+\Delta\rho,\phi) - A_z(\rho-\Delta\rho,\phi)}{2\Delta\rho}$$

$$+ \frac{A_z(\rho,\phi+\Delta\phi) - 2A_z(\rho,\phi) + A_z(\rho,\phi-\Delta\phi)}{(\rho\Delta\phi)^2} = -\mu J_z$$

$$A_z(\rho+\Delta\rho,\phi) - 2A_z(\rho,\phi) + A_z(\rho-\Delta\rho,\phi) + \frac{\Delta\rho}{2\rho}[A_z(\rho+\Delta\rho,\phi) - A_z(\rho-\Delta\rho,\phi)]$$

$$+ \left(\frac{\Delta\rho}{\rho\Delta\phi}\right)^2\Big[A_z(\rho,\phi+\Delta\phi) - 2A_z(\rho,\phi) + A_z(\rho,\phi-\Delta\phi) = -(\Delta\rho)^2\mu J_z$$

If we let $\alpha = \frac{\Delta\rho}{2\rho}$ and $\beta = \left(\frac{\Delta\rho}{\rho\Delta\phi}\right)^2$,

$$A_z(\rho,\phi) = \frac{1}{2+2\beta}\Big[(1+\alpha)A_z(\rho+\Delta\rho,\phi) + (1-\alpha)A_z(\rho-\Delta\rho,\phi) + \beta A_z(\rho,\phi+\Delta\phi)$$

$$+ \beta A_z(\rho,\phi-\Delta\phi) + (\Delta\rho)^2\mu J_z$$

Prob. 3.27 For $\Delta = 0.25$, no. of iterations $= 500$, we apply the ideas in section 3.10 and obtain the following results.

ρ	z	Potential
5	18	65.852
5	10	23.325
5	2	6.3991
10	2	10.226
15	2	10.343

Prob. 3.28 Applying the finite difference scheme of section 3.10, for $\Delta = 0.25$ and 100 iterations, we obtain the following results.

71

ρ	z	Potential
5	18	11.474
5	10	27.870
5	2	12.128
10	2	2.342
15	2	0.3965

Prob. 3.29 For TM ($H_z = 0$) and $\frac{\partial}{\partial\phi} = 0$, $H_\rho = 0 = E_\phi$ and Maxwell's equations become

$$\epsilon\frac{\partial E_\rho}{\partial t} = -\frac{\partial H_\phi}{\partial z} - \sigma E_\rho \tag{1}$$

$$\epsilon\frac{\partial E_z}{\partial t} = \frac{1}{\rho}\frac{\partial}{\partial\rho}(\rho H_\phi) - \sigma E_z = \frac{\partial H_\phi}{\partial\rho} + \frac{H_\phi}{\rho} - \sigma E_z \tag{2}$$

$$\mu\frac{\partial H_\phi}{\partial t} = \frac{\partial E_z}{\partial\rho} - \frac{\partial E_\rho}{\partial z} \tag{3}$$

Applying Eqs. (3.75) and (3.76) to (1) to (3) gives

$$H_\phi^{n+1}(i,j) = H_\phi^n(i,j) + \alpha\left[E_z^{n+1/2}(i,j+1/2) - E_z^{n+1/2}(i,j-1/2)\right]$$
$$- \alpha\left[E_\rho^{n+1/2}(i+1/2,j) - E_\rho^{n+1/2}(i-1/2,j)\right]$$
$$E_z^{n+3/2}(i,j+1/2) = \gamma E_z^{n+1/2}(i,j+1/2) + \beta\left[H_\phi^{n+1}(i,j+1) - H_\phi^{n+1}(i,j)\right.$$
$$\left.+ \frac{1}{j}H_\phi^{n+1}(i,j+1/2)\right]$$
$$E_\rho^{n+3/2}(i+1/2,j) = \gamma E_z^{n+1/2}(i+1/2,j) + \beta\left[H_\phi^{n+1}(i+1,j) - H_\phi^{n+1}(i,j)\right]$$

where $\alpha = \delta t/\mu\delta$, $\beta = \delta t/\epsilon\delta$, $\gamma = 1 - \sigma\delta/\epsilon$.

Prob. 3.30 (b) At $x = 0$,

$$E_z^{n+1}(0,j,k+1/2) = E_z^n(1,j,k+1/2) + \frac{c_o\delta t - \delta}{c_o\delta t + \delta}\left[E_z^{n+1}(1,j,k+1/2) - E_z^{n+1}(0,j,k+1/2)\right]$$

Prob. 3.31

$$\mu_o \frac{\partial H_{xy}}{\partial t} + \sigma_y^* H_{xy} = -\frac{\partial}{\partial y}(E_{zx} + E_{zy}) \tag{1}$$

$$\mu_o \frac{\partial H_{xz}}{\partial t} + \sigma_z^* H_{xz} = \frac{\partial}{\partial z}(E_{yx} + E_{yz}) \tag{2}$$

$$\mu_o \frac{\partial H_{yz}}{\partial t} + \sigma_z^* H_{yz} = -\frac{\partial}{\partial z}(E_{xy} + E_{xz}) \tag{3}$$

$$\mu_o \frac{\partial H_{yx}}{\partial t} + \sigma_x^* H_{yx} = \frac{\partial}{\partial x}(E_{zx} + E_{zy}) \tag{4}$$

$$\mu_o \frac{\partial H_{zx}}{\partial t} + \sigma_x^* H_{zx} = -\frac{\partial}{\partial z}(E_{yx} + E_{yz}) \tag{5}$$

$$\mu_o \frac{\partial H_{zy}}{\partial t} + \sigma_y^* H_{zy} = \frac{\partial}{\partial y}(E_{xy} + E_{xz}) \tag{6}$$

$$\epsilon_o \frac{\partial E_{xy}}{\partial t} + \sigma_y^* E_{xy} = \frac{\partial}{\partial y}(H_{zx} + H_{zy}) \tag{7}$$

$$\epsilon_o \frac{\partial E_{xz}}{\partial t} + \sigma_z^* E_{xz} = -\frac{\partial}{\partial z}(H_{yx} + H_{yz}) \tag{8}$$

$$\epsilon_o \frac{\partial E_{yz}}{\partial t} + \sigma_z^* E_{yz} = -\frac{\partial}{\partial x}(H_{zx} + H_{zy}) \tag{9}$$

$$\epsilon_o \frac{\partial E_{yx}}{\partial t} + \sigma_x^* E_{yx} = -\frac{\partial}{\partial x}(H_{zx} + H_{zy}) \tag{10}$$

$$\epsilon_o \frac{\partial E_{zx}}{\partial t} + \sigma_x^* E_{zx} = \frac{\partial}{\partial x}(H_{yx} + H_{yz}) \tag{11}$$

$$\epsilon_o \frac{\partial E_{zy}}{\partial t} + \sigma_y^* E_{zy} = -\frac{\partial}{\partial y}(H_{xy} + H_{xz}) \tag{12}$$

Prob. 3.32(a) Maxwell's equations can be written as

$$\nabla \times \mathbf{H} = \epsilon \frac{\partial \mathbf{E}}{\partial t} + \sigma E$$

$$\nabla \times \mathbf{E} = -\mu \frac{\partial \mathbf{H}}{\partial t} + \sigma^* H$$

73

In Cartesian coordinates, these become

$$\mu\frac{\partial H_x}{\partial t} = \frac{\partial E_y}{\partial z} - \frac{\partial E_z}{\partial y} - \sigma^* H_x$$

$$\mu\frac{\partial H_y}{\partial t} = \frac{\partial E_z}{\partial x} - \frac{\partial E_x}{\partial z} - \sigma^* H_y$$

$$\mu\frac{\partial H_z}{\partial t} = \frac{\partial E_x}{\partial z} - \frac{\partial E_y}{\partial x} - \sigma^* H_z$$

$$\epsilon\frac{\partial E_x}{\partial t} = \frac{\partial H_z}{\partial y} - \frac{\partial H_y}{\partial z} - \sigma E_x$$

$$\epsilon\frac{\partial E_y}{\partial t} = \frac{\partial H_x}{\partial z} - \frac{\partial H_z}{\partial x} - \sigma E_y$$

$$\epsilon\frac{\partial E_z}{\partial t} = \frac{\partial H_y}{\partial z} - \frac{\partial H_x}{\partial y} - \sigma E_z$$

For $\frac{\partial}{\partial z} = 0 = H_z$, $E_x = 0 = E_y$. Thus for TM case, Maxwell's equation become

$$\epsilon\frac{\partial E_z}{\partial t} + \sigma E_z = \frac{\partial H_y}{\partial z} - \frac{\partial H_x}{\partial y}$$

$$\mu_o\frac{\partial H_x}{\partial t} + \sigma^* H_x = -\frac{\partial E_z}{\partial y}$$

$$\mu_o\frac{\partial H_y}{\partial t} + \sigma^* H_y = \frac{\partial E_z}{\partial x}$$

If we let $E_z = E_{zx} + E_{zy}$, we obtain

$$\epsilon_o\frac{\partial E_{zx}}{\partial t} + \sigma_x E_{zx} = \frac{\partial H_y}{\partial x}$$

$$\epsilon_o\frac{\partial E_{zy}}{\partial t} + \sigma_y E_{zy} = -\frac{\partial H_x}{\partial y}$$

$$\mu_o\frac{\partial H_x}{\partial t} + \sigma_y^* H_x = -\frac{\partial}{\partial y}(E_{zx} + E_{zy})$$

$$\mu_o\frac{\partial H_y}{\partial t} + \sigma_x^* H_y = \frac{\partial}{\partial x}(E_{zx} + E_{zy})$$

(b) The corresponding FDTD equations are

$$E_{zx}^{n+1}(i,j) = e^{-\sigma_x(i)\delta/\epsilon_o} E_{zx}^n(i,j)$$
$$+ \frac{(1 - e^{-\sigma_x(i)\delta/\epsilon_o}}{\sigma_x(i)\delta} \left[H_y^{n+1/2}(i+1/2,j) - H_y^{n+1/2}(i-1/2,j) \right]$$

$$E_{zy}^{n+1}(i,j) = e^{-\sigma_x(i)\delta/\epsilon_o} E_{zy}^n(i,j)$$
$$+ \frac{(1 - e^{-\sigma_x(i)\delta/\epsilon_o}}{\sigma_x(i)\delta} \left[H_x^{n+1/2}(i,j-1/2) - H_x^{n+1/2}(i,j+1/2) \right]$$

$$H_x^{n+1}(i,j+1/2) = e^{-\sigma_y^*(j+1/2)\delta/\mu_o} H_x^{n-1/2}(i,j+1/2)$$
$$+ \frac{(1 - e^{-\sigma_x(j+1/2)\delta/\mu_o}}{\sigma_y^*(j+1/2)\delta} \left[E_{zx}^n(i,j) + E_{zy}^n(i,j) - E_{zx}^n(i,j+1) \right.$$
$$\left. - E_{zy}^n(i,j+1) \right]$$

$$H_y^{n+1/2}(i+1/2,j) = e^{-\sigma_x^*(i+1/2)\delta/\mu_o} H_y^{n-1/2}(i+1/2,j)$$
$$+ \frac{(1 - e^{-\sigma_x(i+1/2)\delta/\mu_o}}{\sigma_x^*(i+1/2)\delta} \left[E_{zx}^n(i+1,j) + E_{zy}^n(i+1,j) - E_{zx}^n(i,j) \right.$$
$$\left. - E_{zy}^n(i,j) \right]$$

Prob. 3.33 Let $H_z = e^{j\omega t}e^{-jk_x x}e^{-jk_z z}$, $E_{yz} = E_{yx} = E_y = e^{j\omega t}e^{-jk_x x}e^{-jk_z z}$. Substituting these into the FDTD equation yields

$$H_z^{n+1/2}(i+1/2,k)\left[e^{j\omega\delta t/2} - e^{-j\omega\delta t/2} \right] = -\frac{\delta t}{\mu\delta} E_y^n(i+1/2,k)\left[e^{-jk_x\delta/2} - e^{jk_x\delta/2} \right]$$

$$j2H_z^n \sin(\omega\delta t/2) = \frac{\delta t}{\mu\delta} E_y^n \sin(k_x\delta/2)$$

or

$$Z_z = \frac{E_y}{H_z} = \frac{\mu_o\delta}{\delta t}\frac{\sin(\omega\delta t/2)}{\sin(k_o\delta/2)}$$

Prob. 3.34

$$H_\rho^{n+1/2}(i,j) = H_\rho^{n-1/2}(i,j) - \frac{m\delta t}{\mu\rho_i} E_z^n(i,j) + \frac{\delta t}{\mu\delta}[E_\phi^n(i,j+1) - E_\phi^n(i,j)]$$

$$H_\phi^{n+1/2}(i,j) = H_\phi^{n-1/2}(i,j) - \frac{\delta t}{\mu\delta}[E_\rho^n(i,j+1) - E_\rho^n(i,j) + E_z^n(i,j) - E_z^n(i+1,j)]$$

Prob. 3.35(a) Exact: $y' = \cos x \Big|_{x=0.44} = 0.9047$

Finite difference:

$$y' = \frac{y(x+\Delta x) - y(x-\Delta x)}{2\Delta} = \frac{y(0.46) - y(0.42)}{2(0.2)}$$
$$= \frac{0.44395 - 0.40776}{0.4} = 0.90475$$

(b) Exact: $\int_{0.4}^{0.52} y\,dx = -\cos x \Big|_{0.4}^{0.52} = 0.05324$

Finite difference: $\Delta x = 0.02$

$$\int_{0.4}^{0.52} y\,dx = \frac{0.02}{3}\big[0.38942 + 4(0.40776) + 2(0.42594) + 4(0.44395)$$
$$+ 2(0.46178) + 4(0.47943) + 0.49688\big]$$
$$= 0.05324$$

Prob. 3.36 Exact: $\int_0^1 f(x)dx = 4x - \frac{x^3}{3}\Big|_0^1 = 3.667$

(a) Trapezoidal rule, $h = 0.2$,

$$I = h\left[\sum_{i=1}^{n-1} f_i + \frac{1}{2}(f_o + f_n)\right]$$
$$= 0.2[f(0.2) + f(0.4) + f(0.6) + f(0.8) + \frac{1}{2}(4 + 3)]$$
$$= 0.2[3.96 + 3.84 + 3.64 + 3.36 + 3.5]$$
$$= 3.66$$

(b) Newton-Cotes with $n = 3$, $f(0) = 4$. Let $m = 2n = 6$, $f(1/6) = 3.96$, $f(2/6) = 3.89$, $f(3/6) = 3.75$, $f(4/6) = 3.56$, $f(5/6) = 3.3$, $f(6/6) = 3$.

$$A_1 = \frac{3(\frac{1}{6})}{8}\left[4 + 3(3.87) + 3(3.89) + 3.75\right] = 1.96$$
$$A_2 = 0.0625[3.75 + 3(3.56) + 3(3.3) + 3] = 1.71$$
$$A = A_1 + A_2 = 3.67$$

Prob. 3.37 The exact solution is found in any standard text on antenna theory, i.e.

$$\int_0^\pi \frac{\cos^2(\frac{\pi}{2}\cos\theta)}{\sin\theta}d\theta = \frac{1}{2}\left[\frac{(2\pi)^2}{2(2!)} - \frac{(2\pi)^4}{4(4!)} + \frac{(2\pi)^6}{6(6!)} + \cdots\right]$$
$$= 1.218$$

Prob. 3.38 Exact: $\int_0^1 e^{-x}dx = -e^{-x}\Big|_0^1 = 1 - 1/e = 0.6321$

Numerical: For $n = 2$, $m = 2$, $h = 0.5$,

$$A = \frac{1}{3}(0.5)[e^0 + 4e^{-0.5} + e^{-1}] = 0.6323$$

For $n = 4$, $m = 4$, $h = 0.25$,

$$A = \frac{4}{90}(0.25)\left[7e^0 + 32e^{-0.25} + 12e^{-0.5} + 32e^{-0.75} + 7e^{-1}\right] = 0.6321$$

For $n = 6$, $m = 6$, $h = 0.167$,

$$A = \frac{6}{840}(0.167)\left[41e^0 + 216e^{-0.167} + 27e^{-0.334} + 272e^{-0.5}\right.$$
$$\left. 27e^{-0.667} + 216e^{-0.834} + 41e^{-1}\right]$$
$$= 0.6320$$

Prob. 3.39 Exact solution: Using the identity $\cos A \sin B = \frac{1}{2}[\sin(A+B) - \sin(A-B)]$

$$\int_0^{2\pi} x\cos 10x \sin 20x\,dx = \frac{1}{2}\int^{2\pi} x\sin 30x\,dx + \frac{1}{2}\int^{2\pi} x\sin 10x\,dx$$
$$= \frac{1}{2}\left[\frac{1}{900}(\sin 30x - 30x\cos 30x) + \frac{1}{100}(\sin 10x - 10x\cos 10x)\right]\Big|_0$$
$$= 0.4333\pi = 1.3613$$

Prob. 3.40 In this case, $n = 3$

$$I = a_o f(x_0) + a_1 f(x_1) + a_2 f(x_2) + a_3 f(x_3)$$

Let $f(x) = 1$,

$$\int_0^1 1\,dx = 4 = a_0 + a_1 + a_2 + a_3 \tag{1}$$

Let $f(x) = x$,

$$\int_0^1 x\,dx = 8 = 0 + a_1 + 3a_2 + 4a_3 \tag{2}$$

Let $f(x) = x^2$,

$$\int_0^1 x^2\,dx = \frac{64}{3} = 0 + a_1 + 9a_2 + 16a_3 \tag{3}$$

Let $f(x) = x^3$,

$$\int_0^1 x^3 dx = 64 = 0 + a_1 + 27a_2 + 64a_3 \qquad (4)$$

Solving (1) to (4) gives

$$a_0 = a_3 = \frac{2}{9}, \quad a_1 = a_2 = \frac{16}{9}$$

Prob. 3.42 Exact solution: $-0.4116 + j0$

Prob. 3.43 Compare your results with the following exact values.

(a) 1.724

(b) 3.963

(c) 15.02

CHAPTER 4

Prob. 4.1(a)

$$< u, v > = \int_{-1}^{1} x^2(2-x)dx = \frac{2x^3}{3} - \frac{x^4}{4}\Big|_{-1}^{1}$$

$$= \frac{4}{3} = 1.333$$

(b)

$$< u, v > = \int_{x=0}^{1} \int_{y=1}^{2} (x^2 - 2y^2)dxdy$$

$$= \frac{x^3}{3}\Big|_{0}^{1}(1) - 2(1)\frac{y^3}{3}\Big|_{1}^{2}$$

$$= -4.667$$

(c)

$$< u, v > = \int \int \int (x+y)xzdxdydz$$

In cylindrical system,

$$x = \rho\cos, \quad y = \rho\sin\phi, \qquad dxdydz = \rho d\phi dzd\rho$$

$$< u, v > = \int_{\rho=0}^{2} \int_{\phi}^{2\pi} \int_{z=0}^{5} (z\rho^2\cos^2\phi + z\rho^2\sin\phi\cos\phi)\rho d\phi dzd\rho$$

$$= \int_{0}^{5} zdz \int_{0}^{2} \rho^3 d\rho \int_{0}^{2\pi} \left[\frac{1}{2}(1+\cos 2\phi) + \frac{1}{2}\sin 2\phi\right]d\phi$$

$$= \frac{25}{2}(4)\pi = 157.08$$

Prob. 4.2(a)

$$< h(-x), f(x) > = \int h(-x)f(x)dx$$

Let $u = -x, \ dx = -du$

$$< h(-x), f(x) > = \int h(u)f(-u)d(-u) = \int h(u)[-f(-x)]du$$

$$= < h(x), -f(-x) >$$

(b)

$$< h(ax), f(x) >= \int h(ax) f(x) dx$$

Let $u = ax$, $dx = du/a$

$$< h(ax), f(x) > = \int h(u) f(u/a) du/a = \int h(u) [\frac{1}{a} f(u/a)] du$$

$$=< h(x), \frac{1}{a} f(\frac{1}{a}) >$$

(c)

$$< f'(x), h(x) >= \int f'(x) h(x) dx$$

Integrating by parts, let $v = h(x)$, $du = f'(x) dx$

$$dv = h'(x) dx, \ u = f(x)$$

$$< f'(x), h(x) >= h(x) f(x) \Big|_a^b - \int f(x) h'(x) dx$$

If $h(x)$ or $f(x)$ vanishes at $x = a$ and $x = b$,

$$< f'(x), h(x) >= - < f(x), h'(x) >$$

(d) By integrating by parts n times as done once in (c), it can readily be shown that

$$< \frac{d^n f}{dx^n}, h(x) >= (-1)^n < f(x), \frac{d^n h}{dx^n} >$$

Prob. 4.3(a) $F(x, y, y') = (1 + y^2)^{1/2}$

$$F_y = 0, \ F'_y = \frac{1}{2}(1 + y')^{-1/2}$$

$$0 = F_y - \frac{d}{dx} F'_y = -\frac{1}{2} \frac{d}{dx} \left[(1 + y')^{-1/2} \right]$$

$$= \frac{1}{4}(1 + y')^{-3/2} y''$$

or

$$y'' = 0$$

(b) $F(x, y, y') = y\sqrt{1 + (y')^2}$

$$0 = F_y - \frac{d}{dx} F'_y = \sqrt{1 + (y')^2} - \frac{d}{dx} \left(\frac{yy'}{\sqrt{1 + (y')^2}} \right)$$

80

or
$$1 + y'^2 - yy'' = 0$$

(c) $F(x, y, y') = \cos(xy')$, $F_y = 0$, $F_y' = -x\sin(xy')$

$$0 = F_y - \frac{d}{dx}F_y' = 0 - \frac{d}{dx}[x\sin(xy')]$$

$$xy'\cos(xy') + \sin(xy') = 0$$

Prob. 4.4(a) $F(x, y, y') = y'^2 - y^2$

$$0 = F_y - \frac{d}{dx}F_y' = -2y - 2y''$$

or
$$y'' + y = 0$$

(b) $F(x, y, y', y'') = 5y^2 - (y'')^2 + 10x$

$$0 = F_y - \frac{d}{dx}F_y' + \frac{d^2}{dx^2}F_{y''} = 10y - 0 + \frac{d^2}{dx^2}(-2y'')$$

or
$$2\frac{d^4y}{dx^4} - 10y = 0$$

(c) $F(x, u, v, u', v') = 3uv - u^2 + u'^2 - v'^2$

$$0 = \frac{\partial F}{\partial u} - \frac{d}{dx}\left(\frac{\partial F}{\partial u_x}\right) = 3v - 2u - \frac{d}{dx}(2u')$$

or
$$3v - 2u - 2u'' = 0$$
$$0 = \frac{\partial F}{\partial v} - \frac{d}{dx}\left(\frac{\partial F}{\partial v_x}\right) = 3u - \frac{d}{dx}(-2v')$$

or
$$3u + 2v'' = 0$$

Prob. 4.5(a) $F(x, y, y') = 2y'^2 + yy' + y' + y$,

$$F_y = y' + 1, \quad F_{y'} = 4y' + y + 1$$

$$0 = F_y - \frac{d}{dx}F_y' = y' + 1 - 4y'' - y' = 0$$

$$y'' = 1/4 \quad \rightarrow \quad y = x^2/8 + Ax + B$$

$y(0) = 0$ gives $B = 0$, while $y(1) = 1$ leads to $A = 7/8$. Thus

$$y = \frac{1}{8}(x^2 + 7x)$$

(b) $F(x, y, y') = y'^2 - y^2$, $F_y = -2y$, $F_{y'} = 2y'$

$$0 = F_y - \frac{d}{dx}F'_y = -2y - 2y'' = 0$$

$$y'' + y = 0 \qquad \rightarrow \qquad y = c_1 \sin x + c_2 \cos x$$

$y(\pi/2) = 0$ and $y(0) = 1$ give $c_1 = 0$ and $c_2 = 1$. Hence

$$y = \cos x$$

Prob. 4.6 Let Φ_o be the unique solution, i.e. $L\Phi_o = g$. Then

$$I = <L\Phi, \Phi> - 2<\Phi, L\Phi_o>$$

Adding and subtracting $<L\Phi_o, \Phi_o>$ and noting that if L is self-adjoint $<L\Phi_o, \Phi> = <L\Phi, \Phi_o>$, we obtain

$$I = <L(\Phi - \Phi_o), \Phi - \Phi_o> - <L\Phi_o, \Phi_o>$$

Since L is positive definite, the last term on the right-hand side is always positive while the first term is greater than or equal to zero. Thus I assumes its least value when $\Phi = \Phi_o$.

Prob. 4.7 $F(x, y, \Phi, \Phi_x, \Phi_y) = \frac{1}{2}(\Phi_x^2 + \Phi_y^2) - \frac{1}{2}k^2\Phi^2 + g\Phi$. Applying Euler's equation

$$\frac{\partial F}{\partial \Phi} - \frac{\partial}{\partial x}\left(\frac{\partial F}{\partial \Phi_x}\right) - \frac{\partial}{\partial y}\left(\frac{\partial F}{\partial \Phi_y}\right) = 0$$

$$-k^2\Phi + g - \frac{\partial}{\partial x}(\Phi_x) - \frac{\partial}{\partial y}(\Phi_y) = 0$$

$$g = \Phi_{xx} + \Phi_{yy} + k^2\Phi$$

or

$$\nabla^2\Phi + k^2\Phi = g$$

Prob. 4.8

$$F(V, V_x, V_y) = |\nabla V|^2 = \left(\frac{\partial V}{\partial x}\right)^2 + \left(\frac{\partial V}{\partial y}\right)^2 = V_x^2 + V_y^2$$

According to Euler's equation

$$0 = \frac{\partial F}{\partial V} - \frac{\partial}{\partial x}\left(\frac{\partial F}{\partial V_x}\right) - \frac{\partial}{\partial y}\left(\frac{\partial F}{\partial V_y}\right)$$

$$= 0 - \frac{\partial}{\partial x}(2V_x) - \frac{\partial}{\partial y}(2V_y)$$

or

$$\frac{\partial^2 V}{\partial x^2} + \frac{\partial^2 V}{\partial y^2} = \nabla^2 V = 0$$

Prob. 4.9

$$F = F(V, V_x, V_y) = \frac{1}{2}\epsilon E^2 - \rho_v V$$

Euler's equation for this 3-D case is

$$\frac{\partial F}{\partial V} - \frac{\partial}{\partial x}\left(\frac{\partial F}{\partial V_x}\right) - \frac{\partial}{\partial y}\left(\frac{\partial F}{\partial V_y}\right) - \frac{\partial}{\partial z}\left(\frac{\partial F}{\partial V_z}\right) \tag{1}$$

But

$$V_x = \frac{\partial V}{\partial x} = -E_x, \quad V_y = \frac{\partial V}{\partial y} = -E_y, \quad V_z = \frac{\partial V}{\partial z} = -E_z$$

$$E = \sqrt{E_x^2 + E_y^2 + E_z^2}, \qquad \frac{\partial F}{\partial V} = -\rho_v$$

Hence

$$\frac{\partial F}{\partial V_x} = -\frac{\partial F}{\partial E_x} = -\frac{\partial F}{\partial E}\frac{\partial E}{\partial E_x} = -\epsilon E_x \tag{2}$$

Similarly,

$$\frac{\partial F}{\partial V_y} = -\epsilon E_y, \qquad \frac{\partial F}{\partial V_z} = -\epsilon E_z \tag{3}$$

Substituting (2) and (3) into (1) gives

$$-\rho_v + \frac{\partial}{\partial x}(\epsilon E_x) + \frac{\partial}{\partial y}(\epsilon E_y) + \frac{\partial}{\partial z}(\epsilon E_z)$$

or

$$\rho_v = \nabla \cdot \epsilon \mathbf{E} = \nabla \cdot \mathbf{D}$$

Prob. 4.10 $F = \frac{1}{2}\sigma E^2 = F(V, V_x, V_y, V_z)$. But

$$\mathbf{E} = -\nabla V = -\frac{\partial V}{\partial x}\mathbf{a}_x - \frac{\partial V}{\partial y}\mathbf{a}_y - \frac{\partial V}{\partial z}\mathbf{a}_z$$

$$E^2 = V_x^2 + V_y^2 + V_z^2$$

According to Euler's equation

$$\frac{\partial F}{\partial V} - \frac{\partial}{\partial x}\left(\frac{\partial F}{\partial V_x}\right) - \frac{\partial}{\partial y}\left(\frac{\partial F}{\partial V_y}\right) - \frac{\partial}{\partial z}\left(\frac{\partial F}{\partial V_z}\right)$$

$$\frac{\partial F}{\partial V_x} = \frac{1}{2}\sigma(2V_x) = \sigma V_x = J_x$$

Hence

$$0 - \frac{\partial}{\partial x}(J_x) - \frac{\partial}{\partial y}(J_y) - \frac{\partial}{\partial z}(J_z) = 0$$

or

$$\nabla \cdot \mathbf{J} = 0$$

Prob. 4.11

$$\delta I = \int\int\int \left[\frac{\partial}{\partial x}\left(\epsilon_x\frac{\partial V}{\partial x}\right) + \frac{\partial}{\partial y}\left(\epsilon_y\frac{\partial V}{\partial y}\right) + \frac{\partial}{\partial z}\left(\epsilon_z\frac{\partial V}{\partial z}\right) + \rho_v\right]\delta V\,dxdydz = 0$$

$$= -\int\int\int \frac{\partial}{\partial x}\left(\epsilon_x\frac{\partial V}{\partial x}\right)\delta V\,dxdydz + \cdots + \int\int\int \rho_v\right]\delta V\,dxdydz = 0$$

To intregrate by parts, let $u = \delta v$, $dv = \frac{\partial}{\partial x}\left(\epsilon_x\frac{\partial V}{\partial x}\right)$, then $du = \frac{\partial}{\partial x}\delta V\,dx$, $v = \epsilon_x\frac{\partial V}{\partial x}$

$$-\int\int\int \frac{\partial}{\partial x}\left(\epsilon_x\frac{\partial V}{\partial x}\right)\delta V\,dxdydz = -\int\int \left[\delta V\epsilon_x\frac{\partial V}{\partial x} - \int \epsilon_x\frac{\partial V}{\partial x}\frac{\partial}{\partial x}\delta V\,dx\right]dxdy$$

Thus

$$\delta I \int\int\int \left[\epsilon_x\frac{\partial V}{\partial x}\frac{\partial}{\partial x}\delta V + \epsilon_y\frac{\partial V}{\partial y}\frac{\partial}{\partial y}\delta V + \epsilon_x\frac{\partial V}{\partial z}\frac{\partial}{\partial z}\delta V + \delta\rho_v V\right]dxdydz$$

$$-\int\int \delta V\epsilon_x\frac{\partial V}{\partial x}dydz - \int\int \delta V\epsilon_y\frac{\partial V}{\partial y}dxdz - \int\int \delta V\epsilon_z\frac{\partial V}{\partial z}dxdy$$

$$= \frac{\delta}{2}\int\int\int \left[\epsilon_x\left(\frac{\partial V}{\partial x}\right)^2 + \epsilon_y\left(\frac{\partial V}{\partial y}\right)^2 + \epsilon_z\left(\frac{\partial V}{\partial z}\right)^2 - 2\rho_v V\right]dxdydz$$

$$= -\delta\int\int V\epsilon_x\frac{\partial V}{\partial x}dydz - \delta\int\int V\epsilon_y\frac{\partial V}{\partial y}dxdz - \delta\int\int V\epsilon_z\frac{\partial V}{\partial z}dxdy$$

The last three terms vanish if we assume either homogeneous Dirichless or Neumann conditions at the boundaries. Thus

$$I = \frac{1}{2}\int_v \left[\epsilon_x\left(\frac{\partial V}{\partial x}\right)^2 + \epsilon_y\left(\frac{\partial V}{\partial y}\right)^2 + \epsilon_z\left(\frac{\partial V}{\partial z}\right)^2 - 2\rho_v V\right]dxdydz$$

Prob. 4.12 Equation (4.24) can be written, for three dimensional problem, as

$$\delta I = \frac{\delta}{2}\int_v (|\nabla\Phi|^2 - 2f\Phi)dv - \delta\int_S \Phi\frac{\partial\Phi}{\partial n}dS$$

84

Substituting $\frac{\partial \Phi}{\partial n} = h - g\Phi$ gives

$$\delta I = \frac{\delta}{2} \int_v \left(|\nabla \Phi|^2 - 2f\Phi\right)dv + \frac{\delta}{2} \int_S \Phi(g\Phi - h)dS$$

or

$$I = \int_v \left(|\nabla \Phi|^2 - 2f\Phi\right)dv + \int_S (g\Phi^2 - 2h\Phi)dS$$

Prob. 4.13 If $-y'' + y = \sin \pi x$,

$$Ly - g = -y'' + y - \sin \pi x = 0$$

$$I = <Ly - g, \delta y> = \int_0^1 (-y'' + y - \sin \pi x)\delta y \, dx$$

$$\delta I = \int_0^1 -y'' \delta y \, dx + \int_0^1 y \delta y \, dx - \int_0^1 \sin \pi x \delta y \, dx$$

$$= -y'\delta y \Big|_0^1 + \frac{\delta}{2}\int_0^1 (y')^2 dx + \frac{\delta}{2}\int y^2 dx + \delta \int y \sin \pi x \, dx$$

$$= \frac{\delta}{2}\int_0^1 \left[y'^2 + y^2 - 2y \sin \pi x\right]dx$$

or

$$I = \frac{1}{2}\int_0^1 \left[(y')^2 + y^2 - 2y \sin \pi x\right]dx$$

Prob. 4.14 $I(\Phi) = <L\Phi, \Phi> - 2<\Phi, g>$, where

$$L = -\frac{d^2}{dx^2} + 1, \ g = 4xe^x$$

$$I(\Phi) = \int \left(-\Phi \frac{d^2\Phi}{dx^2} + \Phi^2\right)dx - 2\int 4xe^x dx$$

Integrating the first term by parts, $u = \Phi, \quad dv = \frac{d^2\Phi}{dx^2}dx$

$$\int \Phi \frac{d^2\Phi}{dx^2}dx = \Phi\frac{d\Phi}{dx}\Big|_0^1 - \int \frac{d\Phi}{dx}\frac{d\Phi}{dx}dx$$

$$I(\Phi) = \int_0^1 \left(\Phi'^2 + \Phi^2 - 8xe^x\Phi\right)dx - \Phi(1)\Phi'(1) + \Phi(0)\Phi'(0)$$

But $\Phi'(1) = \Phi(1) - e$ and $\Phi'(0) = \Phi(0) + 1$,

$$I(\Phi) = \int_0^1 \left(\Phi'^2 + \Phi^2 - 8xe^x\Phi\right)dx - \Phi^2(1) + \Phi^2(0) + e\Phi(1) + \Phi(0)$$

Prob. 4.15 For the exact solution,

$$-\Phi = \frac{x^3}{6} + Ax + B$$

$$\Phi(0) = 0 \quad \rightarrow \quad B = 0$$

$$\Phi(1) = 2 \quad \rightarrow \quad A = -13/6$$

$$\Phi(x) = \frac{13}{6}x - \frac{1}{6}x^3 = 2.1667x - 0.1667x^3$$

If $-\Phi'' = x, \quad \Phi'' + x = 0$

$$\delta I = \int_0^1 (\Phi'' + x)\delta\Phi\, dx$$

$$I = \int_0^1 \left[\frac{1}{2}(\Phi')^2 - x\Phi\right] dx$$

$$F = \frac{1}{2}(\Phi')^2 - x\Phi$$

For $N = 1$,

$$\tilde{\Phi} = u_0 + a_1 u_1 = 2x + a_1(x^2 - x)$$

$$\tilde{\Phi} = 2 + a_1(2x - 1)$$

$$I = \int_0^1 \left(\frac{1}{2}[2 + a_1(2x-1)]^2 - x[2x - a_1(x^2 - x)]\right) dx$$

$$= \frac{4}{3} + \frac{a_1}{12} + \frac{a_1^2}{6}$$

$$\frac{dI}{da_1} = 0 \quad \rightarrow \quad \frac{1}{12} + \frac{a_1}{3} = 0 \quad \rightarrow \quad a_1 = -\frac{1}{4}$$

Hence,

$$\tilde{\Phi} = 2x - \frac{1}{4}x(x-1) = \frac{1}{4}x(9 - x)$$

$$\tilde{\Phi} = 2.25x - 0.25x^2$$

For $N = 2$,

$$\tilde{\Phi} = u_0 + a_1 u_1 + a_2 u_2$$
$$= 2x + a_1(x^2 - x) + a_2(x^3 - x^2)$$

$$I = \int_0^1 [\frac{1}{2}(\Phi')^2 - x\Phi]dx$$

$$= \frac{4}{3} + \frac{a_1}{12} + \frac{a_2}{20} + \frac{a_1 a_2}{6} + \frac{a_1^2}{6} + \frac{a_2^2}{15}$$

$$\frac{\partial I}{\partial a_1} = 0 \qquad \rightarrow \qquad \frac{a_1}{3} + \frac{a_2}{6} = \frac{1}{12}$$

$$\frac{\partial I}{\partial a_2} = 0 \qquad \rightarrow \qquad \frac{a_1}{6} + \frac{2a_2}{15} = -\frac{1}{20}$$

Hence, $a_1 = a_2 = -1/6$ and

$$\tilde{\Phi} = 2x - \frac{1}{6}(x^2 - x) - \frac{1}{6}(x^2 - x^3)$$

$$= 2.1667x - 0.1667x^3$$

which is the same as the exact solution. $N = 3$ gives the same result.

Prob. 4.16(a) $u_m = x(1 - x^m)$. For $m = 1$,

$$\tilde{\Phi} a_1 x(1 - x), \qquad \tilde{\Phi}' = a_1(1 - 2x)$$

Substitution into eq. (4.5.1) in Example 4.5 leads to

$$I(a_1) = \int_0^1 \left[a_1^2(1 - 4x + 4x^2) + 2a_1(x^3 - x^4) - 4a_1^2(x^2 - 2x^3 + x^4) \right] dx$$

$$= \frac{1}{5}a_1^2 + \frac{1}{10}a_1$$

$$\frac{\partial I}{\partial a_1} = \frac{2}{5}a_1 + \frac{1}{10} = 0 \qquad \rightarrow \qquad a_1 = -\frac{1}{4}$$

$$\tilde{\Phi} = a_1 u_1 = -\frac{1}{4}x(1 - x)$$

For $m = 2$,

$$\tilde{\Phi} = a_1 u_1 + a_2 u_2 = a_1(x - x^2) + a_2(x - x^3)$$
$$\tilde{\Phi}' = a_1(1 - 2x) + a_2(1 - 3x^2)$$

$$I(\Phi) = \int_0^1 \left[a_1^2(1 - 4x + 4x^2) + a_2^2(1 - 6x^2 + 9x^4) + a_1 a_2(1 - 2x - 3x^2 + 6x^3) \right.$$
$$- 4a_1^2(x^2 - 2x^3 + x^4) - 4a_2^2(x^2 - 2x^4 + x^6)$$
$$\left. - 4a_1 a_2(x^2 - x^3 - x^4 + x^5) + 2a_1(x^3 - x^4) + 2a_2(x^3 - x^5) \right] dx$$

$$I(a_1, a_2) = \frac{1}{5}a_1^2 + \frac{52}{105}a_1 a_2 + \frac{1}{10}a_1 + \frac{1}{6}a_2$$

$$\frac{\partial I}{\partial a_1} = 0 \quad \rightarrow \quad \frac{2}{5}a_1 + \frac{3}{5}a_2 = -\frac{1}{10}$$

$$\frac{\partial I}{\partial a_2} = 0 \quad \rightarrow \quad \frac{3}{5}a_1 + \frac{104}{105}a_2 = -\frac{1}{6}$$

Solving these gives $a_1 = 1/38$, $a_2 = -7/38$. Hence,

$$\tilde{\Phi} = x(0.0184x^2 - 0.0263x - 0.1579)$$

For $m = 3$,

$$\tilde{\Phi} = a_1(x - x^2) + a_2(x - x^3) + a_3(x - x^4)$$
$$\tilde{\Phi}' = a_1(1 - 2x) + a_2(1 - 3x^2) + a_3(1 - 4x^3)$$

Substituting these in $I(\Phi)$ and taking derivatives, we obtain

$$\frac{\partial I}{\partial a_1} = 0 \quad \rightarrow \quad 0.4a_1 + 0.6a_2 + 0.724a_3 = -0.1$$

$$\frac{\partial I}{\partial a_2} = 0 \quad \rightarrow \quad 0.6a_1 + 0.99a_2 + 1.267a_3 = -1.67$$

$$\frac{\partial I}{\partial a_3} = 0 \quad \rightarrow \quad 0.724a_1 + 1.267a_2 + 1.682a_3 = -0.214$$

Solving these leads to

$$a_1 = -0.223, \quad a_2 = -0.506, \quad a_3 = -0.157$$

(b) $u_m = \sin m\pi x$

$$A_{mn} = \langle u_m, Lu_n \rangle$$
$$= \int_0^1 \sin m\pi x \left(\frac{d^2}{dx^2} + 4 \right) \sin n\pi x\, dx$$
$$= \int_0^1 [4 \sin m\pi x \sin n\pi x - n^2\pi^2 \sin n\pi x \sin m\pi x]\, dx$$
$$= \begin{cases} 0, & m \neq n \\ \dfrac{4 - n^2\pi^2}{2}, & m = n \end{cases}$$

$$B_n = \langle g, u_n \rangle = \int_0^1 x^2 \sin n\pi x\, dx$$
$$= \begin{cases} \dfrac{n^2\pi^2 - 4}{n^3\pi^3}, & n = \text{odd} \\ -\dfrac{1}{n\pi}, & n = \text{even} \end{cases}$$

For $m = 1$,

$$A_{11} = \frac{4 - \pi^2}{2}, \; B_1 = \frac{\pi^2 - 4}{\pi^3} \quad \rightarrow \quad a_1 = B_1/A_{11} = -\frac{3}{\pi^3}$$

For $m = 2$,

$$[A] = \begin{bmatrix} \dfrac{4 - \pi^2}{2} & 0 \\ 0 & \dfrac{4 - 4\pi^2}{2} \end{bmatrix}, \qquad [B] = \begin{bmatrix} \dfrac{\pi^2 - 4}{\pi^3} \\ -\dfrac{1}{2\pi} \end{bmatrix}$$

$$\begin{bmatrix} a_1 \\ a_2 \end{bmatrix} = [A]^{-1}[B] = \begin{bmatrix} -\dfrac{2}{\pi^3} \\ \dfrac{1}{4\pi(\pi^2 - 1)} \end{bmatrix}$$

For $m = 3$,

$$[A] = \begin{bmatrix} \dfrac{4 - \pi^2}{2} & 0 & 0 \\ 0 & 2 - 2\pi^2 & 0 \\ 0 & 0 & 2 - 4.5\pi^2 \end{bmatrix} \qquad [B] = \begin{bmatrix} \dfrac{\pi^2 - 4}{\pi^3} \\ -\dfrac{1}{2\pi} \\ \dfrac{9\pi^2 - 4}{27\pi^3} \end{bmatrix}$$

$$\begin{bmatrix} a_1 \\ a_2 \\ a_3 \end{bmatrix} = [A]^{-1}[B] = \begin{bmatrix} -\dfrac{2}{\pi^3} \\ \dfrac{1}{\pi(\pi^2 - 1)} \\ -\dfrac{2}{27\pi^3} \end{bmatrix}$$

Prob. 4.17(a) For collocation method, select $w_i(x) = \delta(x - x_i)$, where

$$x_i = \frac{10}{6} + \frac{10}{3}(i - 1)$$

(b) For subdomain method, select subdomains $0 \le x \le 10/3$, $10/3 \le x \le 20/3$, $20/3 \le x \le 10$.

(c) For Galerkin method, $w(x) = \tilde{\Phi}$.

(d) For the least squares method, select $w(x) = L\tilde{\Phi}$.

89

The results are as shown below.

Method	a_1	a_2	a_3
(a) Collacation	10.33	−1.46	0.48
(b) Subdomain	10.44	−1.61	0.67
(c) Galerkin	10.21	−1.32	0.35
(d) Least squares	10.21	−1.32	0.35

Prob. 4.18 (a)

$$R = \Phi'' + \Phi + x = (-2 + x - x^2)a_1 + (2 - 6x + x^2 - x^3)a_2 + x$$

$R(1/4) = 0$ and $R(1/2) = 0$ give

$$\begin{bmatrix} 29/16 & -35/64 \\ 7/4 & 7/8 \end{bmatrix} \begin{bmatrix} a_1 \\ a_2 \end{bmatrix} = \begin{bmatrix} 1/4 \\ 1/2 \end{bmatrix}$$

i.e $a_1 = 6/31 = 0.1935$, $a_2 = 40/217 = 0.1843$

(b) $\int_0^1 Rx(1-x)dx = 0$ and $\int_0^1 Rx^2(1-x)dx$ give

$$\begin{bmatrix} 3/10 & 3/20 \\ 3/29 & 13/105 \end{bmatrix} \begin{bmatrix} a_1 \\ a_2 \end{bmatrix} = \begin{bmatrix} 1/12 \\ 1/20 \end{bmatrix}$$

i.e $a_1 = 71/369 = 0.1924$, $a_2 = 7/41 = 0.1707$

(c) $w_1 = \frac{\partial R}{\partial a_1} = -2 + x - x^2$, $w_2 = \frac{\partial R}{\partial a_2} = 2 - 6x + x^2 - x^3$, $\int_0^1 w_1 R dx = 0$ and $\int_0^1 w_2 R dx$ give $a_1 = 1.8754$, $a_2 = 0.1695$

Prob. 4.19 (i)

$$R = L\hat{y} - g = 1 + a_1(-1 - x^4) + a_2(2 - 11x^2 - x^6)$$

Since $\hat{y}$ satisfies the symmetry condition, we consider only the interval $[0, 1]$.

(a) Collocation method: The collocation points are at $x = 1/4$, $3/4$.

$$R(1/4) = 0 \quad \rightarrow \quad \frac{257}{256}a_1 - \frac{5375}{4096}a_2 = 1$$

$$R(3/4) = 0 \quad \rightarrow \quad \frac{337}{256}a_1 + \frac{17881}{4096}a_2 = 1$$

Thus $a_1 = 0.9293$, $a_2 = -0.05115$

(b) Subdomain method: Divide $[0, 1]$ into $[0, 1/2]$ and $[1/2, 1]$.

$$\int w_1 R \quad \rightarrow \quad \frac{81}{100} a_1 - \frac{1453}{2688} a_2 = \frac{1}{2}$$

$$\int w_2 R \quad \rightarrow \quad \frac{111}{160} a_1 + \frac{6317}{2688} a_2 = \frac{1}{2}$$

i.e. $a_1 = 0.9237$, $a_2 = -0.0599$

(c) Galerkin method: $\int_0^1 (1 - x^2) R \, dx = 0$ and $\int_0^1 (x^2 - x^4) R \, dx$ give

$$\frac{76}{105} a_1 + \frac{52}{315} a_2 = \frac{2}{3}$$

$$\frac{62}{315} a_1 + \frac{1325}{3464} a_2 = \frac{2}{15}$$

i.e. $a_1 = 0.9334$, $a_2 = -0.05433$

(d) Least square method: The choice of w_i corresponds to minimizing the mean square residual

$$RE = \frac{1}{2} \int R^2 dx$$
$$= \frac{68}{45} a_1 + \frac{7096}{1155} a_1 a_2 + \frac{63404}{4095} a_2^2 - \frac{12}{5} a_2^2 - \frac{12}{5} a_1 - \frac{76}{21} a_2 + 1$$

$\frac{\partial RE}{\partial a_1} = 0$ and $\frac{\partial RE}{\partial a_2} = 0$ give

$$\frac{68}{45} a_1 + \frac{3548}{1155} a_2 = \frac{6}{5}$$

$$\frac{3548}{1155} a_1 + \frac{63404}{4095} a_2 = \frac{38}{21}$$

i.e. $a_1 = 0.9327$, $a_2 = -0.06818$

(ii) Rayleigh-Ritz method:

$$I(y) = \int_{-1}^1 [y'^2 - (1 + x^2) y^2 - 2y] dx$$

$$\frac{1}{8} I(\hat{y}) = \frac{19}{105} a_1^2 + \frac{78}{945} a_1 a_2 + \frac{331}{3465} a_2^2 - \frac{1}{3} a_1 - \frac{1}{15} a_2$$

$$\frac{\partial I}{\partial a_1} = 0 \quad \rightarrow \quad \frac{38}{105} a_1 + \frac{78}{945} a_2 = \frac{1}{3}$$

$$\frac{\partial I}{\partial a_2} = 0 \quad \rightarrow \quad \frac{78}{105}a_1 + \frac{662}{3465}a_2 = \frac{1}{15}$$

The results are summarized below.

Method	a_1	a_2
Collocation	0.9292	-0.05115
Subdomain	0.9237	-0.05991
Galerkin	0.9334	-0.05433
Least squares	0.9327	-0.06813
Rayleigh-Ritz	0.9334	-0.05433

Prob. 4.20(a) For Galerkin method,

$$a_1 = 0.93344, \ a_2 = 3.6795$$

(b) For the least squares method,

$$a_1 = 0.932718, \ a_2 = 3.6627$$

Prob. 4.21

$$\tilde{\Phi} = 1 - x + a_1 x(1-x) + a_2 x^2(1-x) + a_2 x^3(1-x)$$

$\int_0^1 w_i R\, dx$ gives

$$133a_1 + 63a_2 + 36a_3 = -70$$
$$140a_1 + 108a_2 + 79a_3 = -98$$
$$264a_1 + 252a_2 + 211a_3 = -210$$

Solving these equations yields

$$a_1 = -0.209, \ a_2 = -0.7894, \ a_3 = -0.209$$

and

$$\tilde{\Phi} = (1-x)(1 - 0.209x - 0.789x^2 + 0.209x^3)$$

Prob. 4.22 Let $\tilde{y}(x) = a_1 u_1 + a_2 u_2 + a_3 u_3$ and $u_n = x(1-x^n)$.

$$\tilde{y}(x) = a_1(x - x^2) + a_2(x - x^3) + a_3(x - x^4)$$

$$A_{mn} =< w_m, Lu_n >= \int_0^1 w_m\left[-\frac{d^2}{dx^2}(x-x^{n+1})\right]dx$$

$$[A] = \begin{bmatrix} 0.5 & 0.1875 & 0.0625 \\ 1 & 0.5 & \\ 1.5 & 1.6875 & 1.6875 \end{bmatrix}$$

$$B_{mn} =< w_m, Mu_n >= \int_0^1 w_m(x-x^{n+1})dx$$

$$[B] = \begin{bmatrix} 0.0575 & 0.0615 & 0.0625 \\ 0.2083 & 0.2344 & 0.2438 \\ 0.4219 & 0.4834 & 0.5150 \end{bmatrix}$$

$$[A] = \lambda[B] \qquad \rightarrow \qquad [A][B]^{-1} - \lambda[I] = 0$$

Solving this gives

$$\lambda_1 = 9.968, \ \lambda_2 = 44.9, \ \lambda_3 = 95.6$$

Exact: $\lambda_n = (n\pi)^2$, i.e.

$$\lambda_1 = 9.8696, \ \lambda_2 = 39.48, \ \lambda_3 = 88.83$$

Prob. 4.23

$$\tilde{\lambda}_o = \frac{\int \Phi L\Phi dx}{\Phi^2 dx} = \frac{\int_0^{10} x(x-10)\left[-2+0.1x(x-10)\right]dx}{\int_0^{10} x^2(x-10)^2dx} = 0.2$$

The exact fundamental eigenfunction is $\Psi(x) = \sin(\pi x/10)$ so that

$$\lambda = 0.1 + (\pi/10)^2 = 0.1987$$

Prob. 4.24 The basis functions depend only on ρ. Let

$$u_k = \cos(2k-1)\pi\rho/2$$

such that u_k satisfies the boundary condition $\Phi(\rho=1) = 0$.

$$\tilde{\Phi} = a_1 \cos \pi\rho/2$$

$$\int_S |\nabla\Phi|^2 dS = \int_{\phi=0}^{2\pi}\int_{\rho=1}^1 \left|\frac{d\Phi}{d\rho}\right|^2 \rho d\rho d\phi$$

$$= 2\pi a_1^2 \frac{\pi^2}{4}\int_0^1 \sin^2\frac{2\pi\rho}{2}\rho d\rho$$

$$= \pi\left(\frac{\pi^2}{8}+\frac{1}{4}\right)a_1^2$$

$$\int_S \tilde{\Phi}^2 dS = 2\pi a_1^2 \int_0^1 \cos^2 \frac{2\pi\rho}{2} \rho d\rho = \frac{1}{\pi}\left(\frac{\pi^2}{2} - 1\right)a_1^2$$

$$\lambda = \frac{\int |\tilde{\Phi}|^2 dS}{\tilde{\Phi}^2 dS} = \frac{\pi\left(\frac{\pi^2}{8} + \frac{1}{4}\right)}{\frac{1}{\pi}\left(\frac{\pi^2}{2} - 1\right)} = 5.832$$

compared with the exact value of 5.779.

Prob. 4.25

$$E_y^2 = \sin^2(\pi x/a) + c_1\left[\cos(2\pi x/a) - \cos(4\pi x/a)\right] + c_1^2 \sin^2(3\pi x/a)$$

$$E_y \frac{d^2 E_y}{dx^2} = -\frac{\pi^2}{a^2}\left[\sin^2(\pi x/a) + 5c_1\left[\cos(2\pi x/a) - \cos(4\pi x/a)\right] + 9c_1^2 \sin^2(3\pi x/a)\right]$$

Substituting this in eq. (4.10.2) in Example 4.10 and performing the integration yields

$$\omega^2 \epsilon_o \mu_o\left[(1 + c_1^2) + (\epsilon_r - 1)(\frac{1}{3} + \frac{\sqrt{3}}{2\pi}) + 2c_1(\epsilon_r - 1)(-\frac{\sqrt{3}}{\pi} + \frac{c_1}{3}(\epsilon_r - 1)\right] = \frac{\pi^2}{a^2}(1 + 9c_1)$$

Taking $\epsilon_r = 4$,

$$\frac{4a^2}{\lambda_c^2} = \frac{1 + 9c_1^2}{(2 + 3\sqrt{3}/2\pi) - c_1(\frac{9\sqrt{3}}{2\pi} + c_1^2}$$

Differentiating with respect to c_1 and setting the result equal to zero yields $c_1 = 0.05201$. Substituting this value gives

$$\frac{a}{\lambda_c} = 0.2948$$

Prob. 4.26 Applying the conditions give

$$A = \frac{4}{a^2},\ B = -\frac{6}{a^2},\ C = 0,\ D = 1$$

Hence,

$$H_z = 1 - \frac{6x^2}{a^2} + \frac{4x^3}{a^3}$$

$$H_z^2 = 1 - \frac{12x^2}{a^2} + \frac{8x^3}{a^3} + \frac{36x^4}{a^4} - \frac{48x^5}{a^5} + \frac{16x^6}{a^6}$$

$$\int_0^a H_z^2 dx = 0.4857a$$

$$H_z \frac{d^2 H_z}{dx^2} = -\frac{12}{a^2} + \frac{24x}{a^3} + \frac{72x^2}{a^4} - \frac{192x^3}{a^5} + \frac{96x^4}{a^6}$$

$$\int_0^a H_z \frac{d^2 H_z}{dx^2} dx = -\frac{4.8}{a}$$

Hence,

$$0.4857a \cdot \omega^2 \mu_o \epsilon_o = \frac{4.8}{a}$$

But $\omega = 2\pi f$. Thus

$$4f_c^2 = \frac{4.8c^2}{0.4857\pi^2 a^2}$$

or

$$f_c = \frac{0.50032c}{a}$$

where c is the speed of light in vacuum.

CHAPTER 5

Prob. 5.1(a) Nonsingular, Fredholm integral of the second kind.

$$\frac{5x}{6} + \frac{1}{2}\int_0^1 xt^2 dt = \frac{5x}{6} + \frac{1}{2}x\frac{t^3}{3}\Big|_0^1 = x = \Phi(x)$$

(b) Nonsingular, Volterra integral of the second kind.

$$\cos x - \sin x + 2\int_0^x e^{-t}\sin(x-t)dt = e^{-x} = \Phi(x)$$

(c) Fredholm integral of the second kind.

$$-\cosh x + \frac{\lambda}{[\frac{\lambda}{2}\sinh 2 + \lambda - 1]}\int_{-1}^1 \cosh(x+t)\cosh t\, dt$$

$$= \frac{\cosh x}{[\frac{\lambda}{2}\sinh 2 + \lambda - 1]} = \Phi(x)$$

Prob. 5.2 (a) Note that $\Phi(0) = 5$, $F(x, \Phi) = 2x\Phi(x)$

$$\frac{d\Phi}{dx} = 2x\Phi(x)$$

$$\int \frac{d\Phi}{\Phi} = \int 2x dx \qquad \rightarrow \qquad \ln \Phi = x^2 + \ln c_o$$

$$\Phi = c_o e^{x^2}$$

$$\Phi(0) = 5 \qquad \rightarrow \qquad c_o = 5$$

$$\Phi = 5e^{x^2}$$

(b)

$$\Phi = x - \int_0^x (x-t)\Phi(t)dt$$

$$\Phi' = 1 - \int_0^x \Phi(t)dt, \ \Phi'' = -\Phi$$

$$\Phi(0) = 0, \ \Phi'(0) = 1$$

$$\Phi'' = -\Phi \qquad \rightarrow \qquad \tilde{\Phi} = d_1 \cos x + d_2 \sin x$$

$$\Phi(0) = 0 \qquad \rightarrow \qquad d_1 = 0$$

$$\Phi'(0) = 1 \qquad \rightarrow \qquad d_2 = 1$$

Hence
$$\tilde{\Phi} = \sin x$$

Prob. 5.3 (a) Given $y'' = -y$,

$$y' = -\int_0^x y(t)dt + c_1$$

$$y'(0) = 1 \quad \rightarrow \quad c_1 = 1$$

$$y' = 1 - \int_0^x y(t)dt$$

$$y = x - \int_0^x (x-t)ydt + c_2$$

$$y(0) = 0 \quad \rightarrow \quad c_2 = 0$$

$$y = x - \int_0^x (x-t)y(t)dt$$

(b) If $y'' = -y + \cos x$

$$y' = -\int_0^x y(t)dt + \sin x + c_1$$

$$y'(0) = 1 \quad \rightarrow \quad c_1 = 1$$

$$y = x - \cos - \int_0^x (x-t)y(t)dt + c_2$$

$$y(0) = 0 \quad \rightarrow \quad c_2 = 1$$

$$y = 1 + x - \cos x - \int_0^x (x-t)y(t)dt$$

Prob. 5.4 Assume

$$G = \sum_{n=1}^{\infty} a_n \Phi_n$$

By solving

$$L\Phi_n = \lambda_n \Phi_n$$

gives

$$\Phi_n = \frac{\sqrt{2}}{a} \sin \lambda_n x, \qquad \lambda_n = \frac{n\pi}{a}$$

Hence

$$G(x; x') = \sum_{n=1}^{\infty} \frac{\Phi_n(x)\Phi_n(x')}{\lambda_n^2 - k^2}$$

$$= \frac{2}{a} \sum_{n=1}^{\infty} \frac{\sin \lambda_n x \sin \lambda_n x'}{\lambda_n^2 - k^2}, \qquad \lambda_n = n\pi/a$$

Prob. 5.5 We want to show that

$$(\nabla^2 + k^2)G = \left(\frac{\partial^2}{\partial x^2} + \frac{\partial^2}{\partial z^2} + k^2\right)G = -\delta(x - x')\delta(z - z')$$

$$\left(\frac{\partial^2}{\partial x^2} + \frac{\partial^2}{\partial z^2} + k^2\right)G = \frac{j}{a}\sum\left(-(n\pi/a)^2 - k_n^2 + k^2\right)\sin(m\pi x/a)\cdot$$

$$\sin(n\pi x'/a)\frac{e^{jk_n(z-z')}}{k_n} = 0$$

since $k_n^2 = k^2 - (n\pi/a)^2$.

Prob. 5.6 This is the same as Example 5.5 when $a = b = 1$.

$$\nabla^2 G = \delta(x - x')\delta(y - y') \tag{1}$$

We first determine the eigenfunction of Laplace's equation, i.e.

$$\nabla^2 G = \lambda u$$

where u satisfies the boundary condition.

$$u_{mn} = 2\sin n\pi x \sin n\pi y, \qquad \lambda_{mn} = -(m^2\pi^2 + n^2\pi^2)$$

Thus,

$$G(x, y, x', y') = 2\sum_{m=1}^{\infty}\sum_{n=1}^{\infty} A_{mn}(x', y')\sin n\pi x \sin n\pi y$$

We substitute this in (1). Using orthogonality property and the shifting properties of the delta function

$$-(m^2\pi^2 + n^2\pi^2)A_{mn} = 2\sin n\pi x \sin n\pi y$$

Thus,

$$G(x, y, x', y') = -\frac{4}{\pi^2}\sum_{m=1}^{\infty}\sum_{n=1}^{\infty}\frac{\sin n\pi x \sin n\pi y \sin n\pi x' \sin n\pi y'}{m^2 + n^2}$$

Prob. 5.7 The eigenfunction expansion is obtained easily if the operator L in LG is self-adjoint. To achieve this, let $H = e^x G$. Then

$$LH = H_{xx} + H_{yy} - H = e^x\delta(x - x')\delta(y - y') \text{ on R}$$

$$H = 0 \text{ on S}$$

Let

$$H = \sum\sum c_{mn}\Phi_{mn}$$

where

$$L\Phi_{mn} = \lambda_{mn}\Phi_{mn} \text{ in R}$$

$$\Phi_{mn} = 0 \text{ on S}$$

By the method of separation of variables,

$$\lambda_{mn} = -1 - (m\pi/a)^2 - (n\pi/b)^2$$

$$\Phi_{mn} = \frac{2}{\sqrt{ab}}\sin(m\pi x/a)\sin(n\pi y/b)$$

Multiplying by Φ_{pq} and integrating over R leads to

$$c_{pq}\lambda_{pq} = \frac{2}{\sqrt{ab}}e^{x'}\sin(p\pi x'/a)\sin(q\pi y'/b)$$

Therefore,

$$G(x,y;x',y') = -\frac{4}{ab}e^{x'-x}\sum_{m=1}^{\infty}\sum_{n=1}^{\infty}\frac{[\sin\lambda_a x\sin\lambda_b y\sin\lambda_a x'\sin\lambda_b y']}{[1+\lambda_a^2+\lambda_b^2]}$$

where $\lambda_a = m\pi/a$, $\lambda_b = n\pi/b$.

Prob. 5.8 (a) Sum the series over m

(b) Sum the series over n.

Prob. 5.10 Let

$$G(x,y;x',h_1+h_2) = \sum_{n=1}^{\infty}\Phi_n(y)\sin(n\pi x'/a)\sin(n\pi x/a)$$

where

$$\Phi(y) = \begin{cases} A_n\sinh n\pi y/a, & 0\le y\le h_1 \\ B_n\sin n\pi y/a + C_n\cosh n\pi y/a, & h_1\le y\le h_1+h_2 \\ D_n\sinh[n\pi(b-y)/a], & h_1+h_2\le y\le b \end{cases}$$

G must satisfy

$$\left(\frac{\partial^2}{\partial x^2}+\frac{\partial^2}{\partial y^2}\right)G(x,y;x',y') = -\frac{1}{\epsilon}\delta(x-x')\delta(y-y')$$

and the given boundary and continuity conditions. By imposing those conditions, we obtain

$$A_n = \frac{2\epsilon_{r2}}{n\pi\epsilon_o\Delta_n}\sinh(n\pi h_3/a)$$

$$B_n = \frac{2\epsilon_{r2}}{n\pi\epsilon_o\Delta_n}\sinh(n\pi h_3/a)\left[\epsilon_{r1}\cosh^2(n\pi h_1/a) - \epsilon_{r2}\sinh^2(n\pi h_1/a)\right]$$

$$C_n = \frac{2(\epsilon_{r2}-\epsilon_{r1})}{n\pi\epsilon_o\Delta_n}\sinh(n\pi h_1/a)\cosh^2(n\pi h_1/a)\sinh^2(n\pi h_3/a)$$

$$D_n = \frac{2\eta_n}{n\pi\epsilon_o\Delta_n}$$

where

$$\Delta_n = \epsilon_{r2}\zeta_n \sinh n\pi h_3/a + \epsilon_{r2}\eta_n \cosh n\pi h_3/a$$

$$\zeta_n = \epsilon_{r1} \cosh n\pi h_1/a \cosh n\pi h_2/a + \epsilon_{r2} \sinh n\pi h_1/a \sinh n\pi h_2/a$$

$$\eta_n = \epsilon_{r1} \cosh n\pi h_1/a \sinh n\pi h_2/a + \epsilon_{r2} \sinh n\pi h_1/a \cosh n\pi h_2/a$$

Prob. 5.11

$$\nabla^2 F + k^2 F = \delta(x - x')\delta(y - y')$$

For $\rho > 0$

$$\frac{1}{\rho}\frac{\partial}{\partial \rho}\left(\rho\frac{\partial F}{\partial \rho}\right) + k^2 F = 0$$

or

$$\rho^2 F'' + \rho F' + k^2\rho^2 F = 0$$

$$F(\rho) = AH_0^{(1)}(k\rho) + BH_0^{(2)}(k\rho)$$

Assuming outgoing wave, $B = 0$

$$F(\rho) = A[J_0(k\rho) + jY_0(k\rho)]$$

$$1 = \lim_{\epsilon \to 0} \oint \frac{\partial F}{\partial \rho}dl = \lim_{\epsilon \to 0} \int_0^{2\pi} jA\frac{\partial Y_o(k\rho)}{\partial \rho}\rho d\phi$$

As $\rho \to 0$, $J_0(k\rho) \to 0$, $Y_0(k\rho) = \frac{2}{\pi}\ln k\rho$

$$\lim_{\rho \to 0}\frac{\partial Y_o(k\rho)}{\partial \rho} = \frac{2}{\pi\rho}$$

Hence

$$1 = \int_0^{2\pi} jA\frac{2}{\pi\rho}\rho d\phi = j4A \quad \to \quad A = -\frac{j}{4}$$

$$F(\rho) = -\frac{j}{4}H_0(k\rho)$$

Prob. 5.12 Let

$$h_0^{(2)}(|\mathbf{r} - \mathbf{r}'|) = \begin{cases} \displaystyle\sum_{n=0}^{\infty} c_n h_n^{(2)}(r')j_n(r)P_n(\cos \alpha), & r < r' \\ \displaystyle\sum_{n=0}^{\infty} c_n j_n(r')h_n^{(2)}(r)j_n(r)P_n(\cos \alpha), & r > r' \end{cases} \tag{1}$$

where constants c_n are to be determined. Using the asymptotic formula

$$h_n^{(2)}(z) = \frac{j^{n+1}}{z}e^{-jz}$$

the left-hand side of (1) becomes

$$h_o^{(2)}(|\mathbf{r} - \mathbf{r}'|) \qquad \to_{r' \to \infty, \; \theta' \to 0} \qquad \frac{je^{-jr'}}{r'} e^{jr \cos \theta} \qquad (2)$$

and the right-hand side becomes

$$\to_{r' \to \infty, \; \theta' \to 0} \qquad \frac{je^{-jr'}}{r'} \sum_{n=0}^{\infty} c_n j_n(r) P_n(\cos \theta) \qquad (3)$$

Comparing (2) and (3) with eq. (2.8.5), for a plane wave, shows that $c_n = 2n + 1$. Thus

$$h_0^{(2)}(|\mathbf{r} - \mathbf{r}'|) = \begin{cases} \displaystyle\sum_{n=0}^{\infty} (2n+1) h_n^{(2)}(r') j_n(r) P_n(\cos \alpha), & r < r' \\ \displaystyle\sum_{n=0}^{\infty} (2n+1) j_n(r') h_n^{(2)}(r) j_n(r) P_n(\cos \alpha), & r > r' \end{cases}$$

Prob. 5.13 It is evident that $K(0, y) = 0 = K(1, y)$ and this is satisfied by $\sin \alpha x$.

$$\sin \alpha 1 = \sin n\pi \qquad \to \qquad \alpha = n\pi$$

i.e. $\sin n\pi x$ is a possible solution. Similarly, $K(x, 0) = 0 = K(x, 1)$ leads to $\sin n\pi y$. Hence

$$K(x, y) = \sum_{n=1}^{\infty} A_n \sin n\pi x \sin n\pi y$$

$$(1 - x)y = \sum_{n=1}^{\infty} A_n \sin n\pi x \sin n\pi y, \; 0 < y < x < 1$$

Multiply both sides by $\sin m\pi x \sin m\pi y$ gives

$$\int_{x=0}^{1} \int_{y=0}^{x} (1 - x)y \sin m\pi x \sin m\pi y \, dx \, dy$$

$$= \sum_{n=1}^{\infty} \int_{x=0}^{1} \int_{y=0}^{x} A_n \sin n\pi x \sin n\pi y \sin m\pi x \sin m\pi y \, dx \, dy$$

$$\int_{x=0}^{1} (1 - x) \sin m\pi x \left(-\frac{\cos m\pi y}{m\pi} \right) \Big|_{0}^{x} dx = A_n \frac{1}{2} \frac{1}{2}, \qquad m = n$$

$$\frac{1}{2n^2 \pi^2} = A_n \frac{1}{2} \frac{1}{2}$$

so that

$$A_n = \frac{2}{n^2\pi^2}$$

Hence

$$K(x,y) = 2\sum_{n=1}^{\infty}\frac{\sin n\pi x \sin n\pi y}{n^2\pi^2}$$

If we let $x = 1/2 = y$,

$$\sin n\pi/2 = \begin{cases} 1, & n = \text{even} \\ -1, & n = \text{odd} \end{cases}$$

$$\sin n\pi/2 \sin n\pi/2 = 1 \qquad \text{for all } n$$

$$K(1/2, 1/2) = 2\sum_{n=1}^{\infty}\frac{\sin n\pi/2 \sin n\pi/2}{n^2\pi^2} = 2\frac{1}{n^2\pi^2}$$

But $K(1/2, 1/2) = (1/2)(1/2) = 1/4$

$$1/4 = 2\frac{1}{n^2\pi^2}$$

$$\frac{\pi^2}{8} = \sum_{n=1}^{\infty}\frac{1}{n^2}$$

Prob. 5.14 We first obtain the Green's function using the method of images, i.e. by means of using three suitably placed charges as shown below.

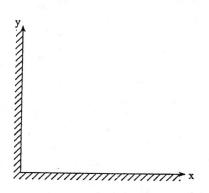

$$\nabla^2 V = g$$
$$\nabla^2 G = \delta(x - x', y - y')$$

$$G(x, y; x', y') = \frac{1}{4\pi}\Big[\ln[(x - x')^2 + (y - y')^2] + \ln[(x + x')^2 + (y - y')^2]$$

$$- \ln[(x - x')^2 + (y + y')^2] - \ln[(x + x')^2 + (y + y')^2]\Big]$$

$$G(x, y; x', y') = \frac{1}{4\pi}\ln\frac{[(x - x')^2 + (y - y')^2][(x + x')^2 + (y - y')^2]}{[(x - x')^2 + (y + y')^2][(x + x')^2 + (y + y')^2]}$$

The solution is given by

$$V(x, y) = \int_L V\frac{\partial G}{\partial n}dl + \int_L G\frac{\partial V}{\partial n}dl + \int_S GgdS$$

$$= \int_o^\infty f\frac{\partial G}{\partial n}dx' + \int_0^\infty Ghdy' + \int_0^\infty \int_0^\infty Ggdx'dy'$$

$$= \int_o^\infty f(x)\frac{\partial G(x, y; x', 0)}{\partial n}dx' + \int_0^\infty G(x, y; 0, y')h(y')dy'$$

$$+ \int_0^\infty \int_0^\infty G(x, y; x', y')g(x', y')dx'dy'$$

Prob. 5.15 From [85],

(a) $Z_o = 62.71 \ \Omega$

(b) $Z_o = 26.75 \ \Omega$

Prob. 5.16 From [86],

(a) $C \simeq 3\epsilon(\text{area}) = 3 \times 9.6 \times \frac{10^{-9}}{36\pi} \times 4 \times 10^{-4} = 0.1019 \ \text{pF}$

(b) $C \simeq 4\epsilon(\text{area}) = 4 \times 9.6 \times \frac{10^{-9}}{36\pi} \times 2 \times 10^{-4} = 0.0.0679 \ \text{pF}$

Prob. 5.17 Apply the techniques discussed in section 5.5.1.

Prob. 5.18 The scatttering patterns are shown below [34].

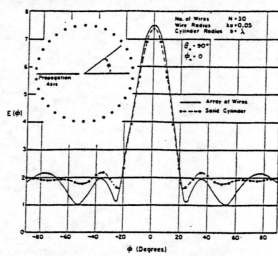

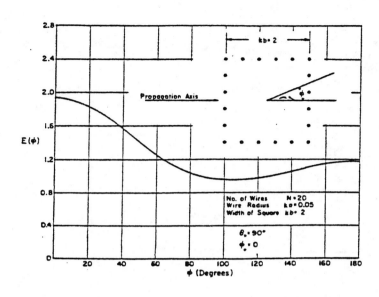

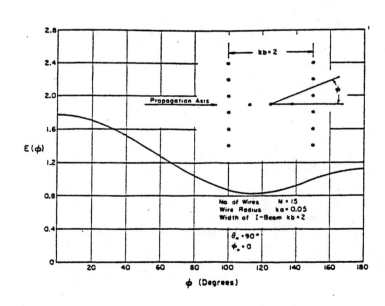

Prob. 5.19 Compare your results with those of the previous problem.

Prob. 5.20 See the figure below.

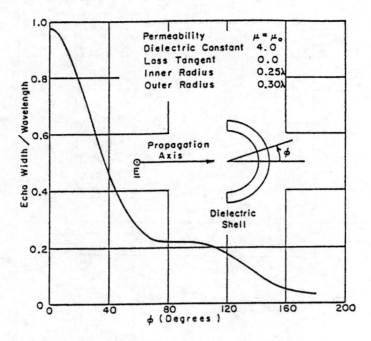

Prob. 5.21 (a) Multiply the given IE by a pulse function

$$u_n(z) = \begin{cases} 1, & z_n - \Delta/2 < z < z_n + \Delta/2 \\ 0, & \text{otherwise} \end{cases}$$

and let

$$I(z) \simeq \sum_{n=1}^{N} I_n u_n(z)$$

We obtain

$$\sum_{n=1}^{N} I_n S_n(z) \simeq f(z)$$

where

$$S_n = -\frac{1}{2\pi} \int_{z_n-\Delta/2}^{z_n+\Delta/2} \ln(z - z') dz'$$

Taking the inner product with

$$w_n(z) = \delta(z - z_n)$$

gives

$$\sum_{n=1}^{N} I_n S_{mn}(z) = F_m$$

or

$$[S][I] = [F]$$

where $F_n = f(z_n)$

$$S_{mn} = S_n(z_m) = -\frac{1}{2\pi} \int_{z_n - \Delta/2}^{z_n + \Delta/2} \ln(z_m - z')dz'$$

But

$$z_n + \Delta/2 = -w + \Delta n - \Delta/2 + \Delta/2 = \Delta n - w$$
$$z_n - \Delta/2 z_n = -w + \Delta n - \Delta/2 - \Delta/2 = \Delta(n-1) - w$$

$$\int \ln x\, dx = x \ln x - x + C$$

$$S_{mn} = -\frac{1}{2\pi} \left[-(z_m - z') \ln(z_m - z') + (z_m - z') \right]_{\Delta n - \Delta - w}^{\Delta n - w}$$

Since

$$z_m - (\Delta n - w) = \Delta(m - n - 1/2)$$
$$z_m - (\Delta n - w - \Delta) = \Delta(m - n + 1/2)$$

$$S_{mn} = -\frac{1}{2\pi} \left[-\Delta(m - n - 1/2) \ln \Delta(m - n - 1/2) + \Delta(m - n - 1/2) \right.$$
$$\left. + \Delta(m - n + 1/2) \ln \Delta(m - n + 1/2) - \Delta(m - n + 1/2) \right]$$

$$S_{mn} = -\frac{1}{2\pi} \left[(m - n) \ln\left(\frac{m - n + 1/2}{m - n - 1/2}\right) + \frac{1}{2} \ln \Delta(m - n + 1/2)(m - n - 1/2) \right]$$
$$= -\frac{1}{2\pi} \left[(m - n) \ln\left(\frac{m - n + 1/2}{m - n - 1/2}\right) + \frac{1}{2} \ln[(m - n)^2 - 1/4] \right.$$
$$+ \frac{1}{2} \ln \Delta + \frac{1}{2} \ln \Delta - 1 \left. \right]$$
$$= \frac{1}{2\pi} \left[1 - \ln \Delta - \frac{1}{2} \ln[(m - n)^2 - 1/4] - (m - n) \ln\left(\frac{m - n + 1/2}{m - n - 1/2}\right) \right.$$
$$+ \frac{1}{2} \ln \Delta(m - n + 1/2)(m - n - 1/2) \left. \right]$$

(b) $[I_m]$ is sketched below.

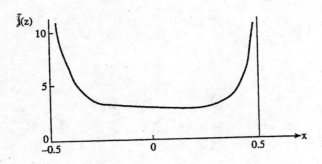

(c) For this case, $[I_m]$ is sketched below.

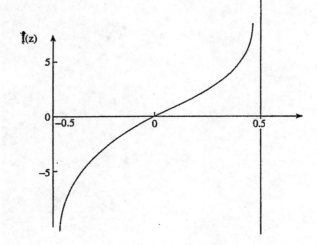

Prob. 5.22 The 3×3 matrix equation is [31]

$$\begin{bmatrix} (6.883, -1.851) & (2.807, -1.768) & (0, 0.0333) \\ (4.894, -1.678) & (5.951, -1.609) & (0, 0.0236) \\ (0.665, -1.218) & (1.129, -1.188) & (0, 0) \end{bmatrix} \begin{bmatrix} B_1 \\ B_2 \\ B_3 \end{bmatrix} = \begin{bmatrix} (0, 0) \\ (0, -0.0118) \\ (0, -0.01667) \end{bmatrix}$$

Solving this matrix equation gives

$$B_1 = (0.0094, -0.00357), \qquad B_2 = (0.00045, -0.00203)$$

These values of B_1 and B_2 are substituted in

$$I(z) = \sum_{n=1}^{2} B_n \sin\left(\frac{2\pi n}{\lambda}\left[\frac{\lambda}{4} - |z|\right]\right)$$

The real and imaginary parts of $I(z)$ are sketched below.

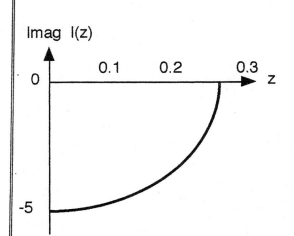

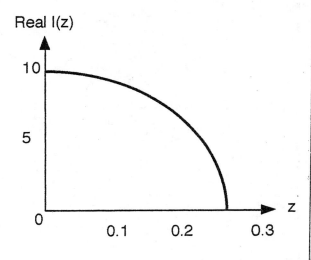

Prob. 5.23 (a) $I(z)$ is as sketched below.

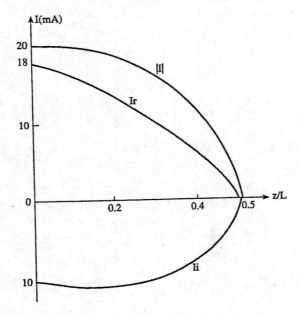

(b) For this case, $I(z)$ is as sketched below.

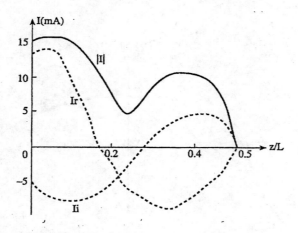

Prob. 5.24 (a) From eq. (5.102),

$$-E_z = j\omega\left(1 + \frac{1}{k^2}\frac{\partial^2}{\partial z^2}\right)\mu \int_{-l/2}^{l/2} I(z')\frac{e^{-jkR}}{4\pi R}dz'$$

$$= \frac{j\omega\mu}{4\pi k^2}\int_{-l/2}^{l/2} I(z')\left(1 + \frac{1}{k^2}\frac{\partial^2}{\partial z^2}\right)\frac{e^{-jkR}}{R}dz' \qquad (1)$$

Now

$$\frac{j\omega\mu}{4\pi k^2} == \frac{j\omega\mu}{4\pi\frac{2\pi}{\lambda}\omega\sqrt{\mu\epsilon}} = \frac{j\lambda\sqrt{\mu/\epsilon}}{8\pi^2} \tag{2}$$

$$R = \left[(x-x')^2 + (y-y')^2 + (z-z')^2\right]^{1/2}$$

$$\frac{\partial}{\partial z}R = \frac{z-z'}{R}, \qquad \frac{\partial}{\partial z}\left(\frac{1}{R}\right) = -\frac{z-z'}{R^3}$$

$$\frac{\partial}{\partial z}\left(\frac{1}{R^3}\right) = -\frac{3(z-z')}{R^5}$$

$$\frac{\partial}{\partial z}\left(\frac{e^{-jkR}}{R}\right) = -e^{-jkR}\frac{(z-z')}{R^3} - \frac{jk(z-z')}{R}\frac{e^{-jkR}}{R}$$

$$= -\frac{(z-z')}{R^2}\frac{e^{-jkR}}{R}(1+jkR)$$

$$\frac{\partial^2}{\partial z^2}\left(\frac{e^{-jkR}}{R}\right) = -\frac{e^{-jkR}}{R^3}(1+jkR) + 3\frac{e^{-jkR}}{R^5}(z-z')^2(1+jkR) - jk(z-z')^2\frac{e^{-jkR}}{R^4}$$

$$+ jk(z-z')(1+jkR)\frac{e^{-jkR}}{R^4}$$

$$= -\frac{e^{-jkR}}{R^3}(1+jkR) + 3\frac{e^{-jkR}}{R^3}(z-z')^2(1+jkR) - jk^2(z-z')^2\frac{e^{-jkR}}{R^3}$$

But $(z-z')^2 = R^2 - (x-x')^2 - (y-y')^2 = R^2 - a^2$. Hence

$$\left(\frac{\partial^2}{\partial z^2} + k^2\right)\frac{e^{-jkR}}{R} = \frac{e^{-jkR}}{R^5}\left[k^2R^4 - R^2(1+jkR) - k^2(R^2-a^2)R^2\right.$$

$$\left. + 3(R^2-a^2)(1+jkR)\right]$$

$$= \frac{e^{-jkR}}{R^5}\left[k^2a^2R^2 + (1+jkR)(2R^2 - 3a^2)\right] \tag{3}$$

Substituting (2) and (3) into (1) gives

$$-E_z = \frac{j\lambda\sqrt{\mu/\epsilon}}{8\pi^2}\int_{-l/2}^{l/2}I(z')\frac{e^{-jkR}}{R^5}\left[k^2a^2R^2 + (1+jkR)(2R^2 - 3a^2)\right]dz'$$

(b) Let $z' = z + a\tan\theta'$, $dz' = a\sec^2\theta'd\theta'$,

$$z' = -l/2 \qquad \rightarrow \qquad \theta_1 = -\tan^{-1}\left(\frac{l/2 + z}{a}\right)$$

$$z' = l/2 \qquad \rightarrow \qquad \theta_2 = -\tan^{-1}\left(\frac{l/2 - z}{a}\right)$$

$$R = \sqrt{a^2 + (z - z')^2} = a \sec \theta'$$

$$e^{-jkR} = e^{-jka \sec \theta'} = e^{-jka/ \cos \theta}$$

$$\frac{1}{R^5}\left[(1 + jkR)(2R^2 - 3a^2) + k^2a^2R^2\right]dz'$$
$$= \frac{1}{a^5 \sec^5 \theta'}\left[(1 + jka \sec \theta')(2a^2 \sec^2 \theta'^2 - 3a^2) + k^2a^4 \sec^2 \theta\right] a \sec^2 \theta' d\theta'$$
$$= \frac{1}{a^2}\left[(jka + \cos \theta')(2 - 3\cos^2 \theta') + k^2a^2 \cos \theta'\right] d\theta'$$

Thus

$$-E_z^i = \frac{\lambda\sqrt{\mu/\epsilon}}{8j\pi^2a^2}\int_{\theta_1}^{\theta_2} I(\theta')e^{-jka/cos\theta'}\left[(jka + \cos \theta')(2 - 3\cos \theta') + k^2a^2 \cos \theta'\right] d\theta$$

Prob. 5.25 The distribution of the normalized induced field $|E_x|/|E^i|$ is shown below [51].

0.0159	0.0161	0.0155	0.21
0.011	0.112	0.108	0.0155
0.0115	0.0116	0.0112	0.016
0.0112	0.0114	0.011	0.0158
0.0112	0.0114	0.011	0.0158
0.0112	0.0114	0.011	0.0158
0.0112	0.0114	0.011	0.0158
0.0112	0.0114	0.011	0.0158
0.0112	0.0114	0.011	0.0158
0.0115	0.0116	0.0112	0.0155
0.0110	0.0112	0.0108	0.0155
0.0159	0.0161	0.0155	0.0211

Prob. 5.26 The absorbed power density calculated at the center of each cell for the first and second layers is shown below [56].

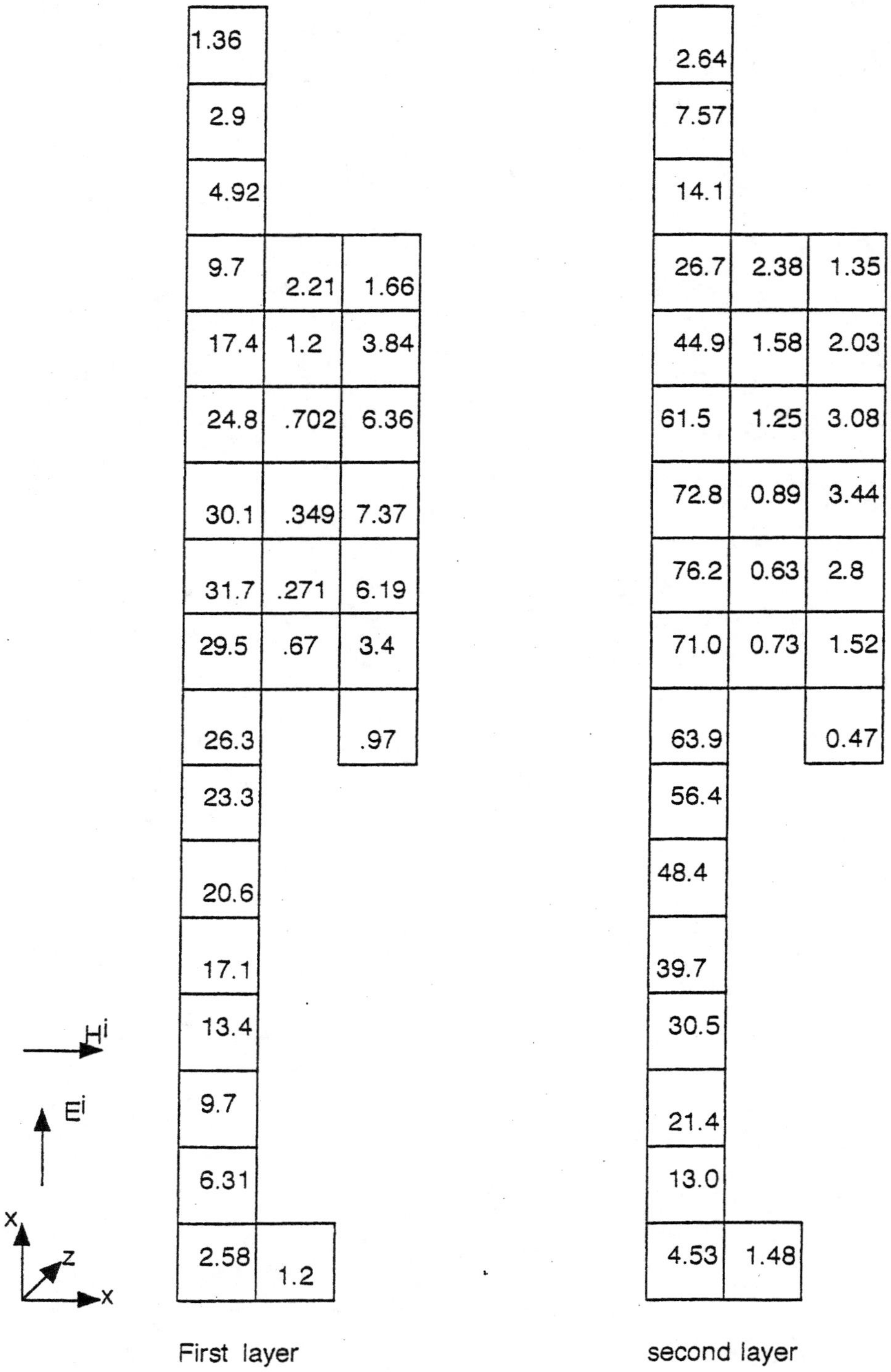

First layer (left):

1.36		
2.9		
4.92		
9.7	2.21	1.66
17.4	1.2	3.84
24.8	.702	6.36
30.1	.349	7.37
31.7	.271	6.19
29.5	.67	3.4
26.3		.97
23.3		
20.6		
17.1		
13.4		
9.7		
6.31		
2.58	1.2	

H^i →

E^i ↑

Second layer (right):

2.64		
7.57		
14.1		
26.7	2.38	1.35
44.9	1.58	2.03
61.5	1.25	3.08
72.8	0.89	3.44
76.2	0.63	2.8
71.0	0.73	1.52
63.9		0.47
56.4		
48.4		
39.7		
30.5		
21.4		
13.0		
4.53	1.48	

First layer   second layer

Prob. 5.27 See the results below.

Cell No.	E_n
64	0.1342
65	0.3966
66	0.4292
67	0.1749
74	0.0965
75	0.3925
76	0.4173
77	0.1393
84	0.1342
85	0.3965
86	0.4293
87	0.1749

CHAPTER 6

Prob. 6.1 (a)

$$P_1 = 1.5, \ P_2 = 0.5, \ P_3 = -2, \ Q_1 = 1, \ Q_2 = 1.5, \ Q_3 = -0.5$$

$$A = \frac{1}{2}(P_2 Q_2 - P_3 Q_2) = 1.375$$

$$C_{ij} = \frac{1}{4A}[P_i P_j + Q_i Q_j]$$

$$C = \begin{bmatrix} 0.5909 & -0.1364 & -0.4545 \\ -0.1304 & 0.4545 & -0.3182 \\ -0.4545 & -0.3182 & 0.7727 \end{bmatrix}$$

(b)

$$P_1 = -4, \ P_2 = 4, \ P_3 = 0, \ Q_1 = 0, \ Q_2 = -3, \ Q_3 = 3$$

$$A = \frac{1}{2}(P_2 Q_2 - P_3 Q_2) = 6$$

$$C = \begin{bmatrix} 0.6667 & -0.6667 & 0 \\ -0.6667 & 1.042 & -0.375 \\ 0 & -0.375 & 0.375 \end{bmatrix}$$

Prob. 6.2 For element 1,

$$P_1 = -0.5, \ P_2 = -1, \ P_3 = 1.5, \ Q_1 = 1, \ Q_2 = -1.2, \ Q_3 = 0.2, \ A = 0.8$$

$$C^{(1)} = \begin{bmatrix} 0.3906 & -0.2188 & -0.1719 \\ -0.2188 & 0.7623 & -0.5438 \\ -0.1719 & -0.5438 & 0.7156 \end{bmatrix}$$

For element 2,

$$P_1 = 1, \ P_2 = -1.5, \ P_3 = 0.5, \ Q_1 = 1.2, \ Q_2 = 0.5, \ Q_3 = -1.7, \ A = 1.15$$

$$C^{(2)} = \begin{bmatrix} 0.5304 & -0.1957 & -0.3348 \\ -0.1957 & 0.5435 & -0.3478 \\ -0.3348 & -0.3478 & 0.6826 \end{bmatrix}$$

The global matrix is

$$C = \begin{bmatrix} C_{22}^{(1)} & C_{23}^{(1)} & 0 & C_{12}^{(1)} \\ C_{23}^{(1)} & C_{33}^{(1)} + C_{33}^{(2)} & C_{13}^{(2)} & C_{13}^{(1)} + C_{23}^{(2)} \\ 0 & C_{12}^{(2)} & C_{11}^{(2)} & C_{12}^{(2)} \\ C_{12}^{(1)} & C_{12}^{(1)} + C_{23}^{(2)} & C_{12}^{(2)} & C_{11}^{(1)} + C_{22}^{(2)} \end{bmatrix}$$

$$= \begin{bmatrix} 0.763 & -0.544 & 0 & -0.219 \\ -0.544 & 1.398 & -0.335 & -0.520 \\ 0 & -0.335 & 0.53 & -0.196 \\ -0.219 & -0.520 & -0.196 & 0.934 \end{bmatrix}$$

Since $C_{ff}V_f = -C_{fp}V_p$,

$$\begin{bmatrix} C_{11} & C_{13} \\ C_{31} & C_{33} \end{bmatrix} \begin{bmatrix} V_1 \\ V_3 \end{bmatrix} = - \begin{bmatrix} C_{11} & C_{14} \\ C_{32} & C_{34} \end{bmatrix} \begin{bmatrix} V_1 \\ V_2 \end{bmatrix}$$

or

$$\begin{bmatrix} 0.7625 & 0 \\ 0 & 0.530 \end{bmatrix} \begin{bmatrix} V_1 \\ V_3 \end{bmatrix} = \begin{bmatrix} -0.5438 & -0.2188 \\ -0.3348 & -0.1957 \end{bmatrix} \begin{bmatrix} 10 \\ -10 \end{bmatrix} = \begin{bmatrix} 3.25 \\ 1.391 \end{bmatrix}$$

Hence

$$V_1 = 4.26, \ V_3 = 2.62$$

Prob. 6.3

$$A = \frac{1}{2}(P_2 Q_3 - P_3 Q_2) = \frac{1}{2}[(-5)(-5) - 1(2)] = 23/2$$

$$\alpha_1 = \frac{1}{2A}\left[(x_2 y_3 - x_3 y_2) + (y_2 - y_3)x + (x_3 - x_2)y\right]$$

$$= \frac{1}{23}[(0 - 24) + (4 - 0)x + 3y]$$

$$= \frac{1}{23}(4x + 3y - 24)$$

$$\alpha_2 = \frac{1}{2A}\left[(x_3 y_1 - x_1 y_3) + (y_3 - y_1)x + (x_1 - x_3)y\right]$$

$$= \frac{1}{23}[(30 - 0) + (0 - 5)x + 2y]$$

$$= \frac{1}{23}(-5x + 2y + 30)$$

$$\alpha_3 = \frac{1}{2A}\left[(x_1 y_2 - x_2 y_1) + (y_1 - y_2)x + (x_2 - x_1)y\right]$$

$$= \frac{1}{23}[(32 - 15) + x + (-5)y]$$

$$= \frac{1}{23}(x - 5y + 17)$$

Note that $\sum \alpha_i = 1$.

Prob. 6.4

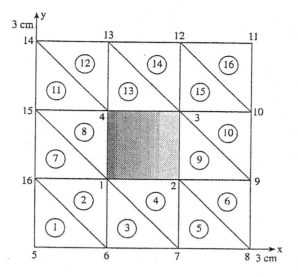

$$C_{3,10} = C_{1,3}^{(10)} + C_{1,2}^{(15)}$$

For element 10,

$$P_1 = 0, \; P_2 = 1, \; P_3 = -1, \; Q_1 = -1, \; Q_2 = 1, \; Q_3 = 0, \; A = 0.5$$

$$C_{1,3}^{(10)} = \frac{1}{2}[(-1)(1) + (0)(1)] = -\frac{1}{2}$$

For element 15,

$$P_1 = -1, \; P_2 = 1, \; P_3 = 0, \; Q_1 = -1, \; Q_2 = 0, \; Q_3 = 1, \; A = 0.5$$

$$C_{1,2}^{(15)} = \frac{1}{2}[(-1)(0) + (0)(-1)] = 0$$

$$C_{3,10} = 0 - \frac{1}{2} = -\frac{1}{2}$$

$$C_{33} = C_{11}^{(9)} + C_{11}^{(10)} + C_{11}^{(13)} + C_{14}^{(5)} + C_{15}^{(5)}$$

For element 9,

$$P_1 = -1, \; P_2 = 1, \; P_3 = 0, \; Q_1 = -1, \; Q_2 = 0, \; Q_3 = 1$$

For element 13,

$$P_1 = -1, \; P_2 = 1, \; P_3 = 0, \; Q_1 = -1, \; Q_2 = 0, \; Q_3 = 1$$

For element 14,

$$P_1 = 0, \ P_2 = 1, \ P_3 = -1, \ Q_1 = -1, \ Q_2 = 1 \ Q_3 =$$

$$C_{3,3} = \frac{1}{2}(0^2 + 1^2) + \frac{1}{2}[(-1)^2 + 0^2] + \frac{1}{2}(1^2 + 0^2) + \frac{1}{2}[0^2 + (-1)^2] + \frac{1}{2}[(-1)^2 + (-1)^2]$$
$$= 3$$

Prob. 6.5

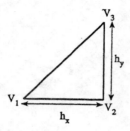

$$W_e = \frac{1}{2}\epsilon[V_e^t][C^{(e)}][V_e], \qquad [V_e] = \begin{bmatrix} V_1 \\ V_2 \\ V_3 \end{bmatrix}$$

$$P_1 = -h_y, \ P_2 = 0, \ P_3 = -h_x, \ Q_1 = 0, \ Q_2 = -h_x, \ Q_3 = h_x$$

$$A = \frac{1}{2}(P_2 Q_3 - P_3 Q_2) = \frac{1}{2}(h_x h_y - 0), \qquad 4A = 2h_x h_y$$

$$C_{ij} = \frac{1}{4A}[P_i P_j + Q_i Q_j]$$

$$C_{11} = \frac{1}{2h_x h_y}[h_y^2 + 0] = \frac{h_y}{2h_x}$$

$$C_{12} = \frac{1}{2h_x h_y}[-h_y^2 + 0] = -\frac{h_y}{2h_x}$$

$$C_{13} = \frac{1}{2h_x h_y}[0 + 0] = 0$$

$$C_{22} = \frac{1}{2h_x h_y}[h_y^2 + h_x^2]$$

$$C_{23} = \frac{1}{2h_x h_y}[0 - h_x^2] = -\frac{h_x}{2h_y}$$

$$C_{33} = \frac{1}{2h_x h_y}[0 - h_x^2] = -\frac{h_x}{2h_y}$$

$$C^{(e)} = \begin{bmatrix} \frac{h_y}{2h_x} & -\frac{h_y}{2h_x} & 0 \\ -\frac{h_y}{2h_x} & \frac{h_y^2+h_x^2}{2h_xh_y} & -\frac{h_x}{2h_y} \\ 0 & -\frac{h_x}{2h_y} & \frac{h_x}{2h_y} \end{bmatrix}$$

$$W_e = \frac{1}{2}\epsilon \begin{bmatrix} V_1 & V_2 & V_3 \end{bmatrix} [C^{(e)}] \begin{bmatrix} V_1 \\ V_2 \\ V_3 \end{bmatrix}$$

$$W_e = \frac{\epsilon}{4h_xh_y}\Big[h_y^2 V_1^2 - 2h_y^2 V_1 V_2 + h_x^2 V_3^2 - h_x^2 V_1 V_3$$

$$+ h_x^2 V_2^2 + h_y^2 V_2^2 - h_x^2 V_2 V_3 \Big]$$

Prob. 6.6

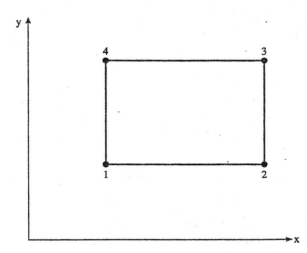

$$V(x,y) = a + bx + cy + dxy$$
$$V_1 = a + bx_1 + cy_1 + dx_1y_1$$
$$V_2 = a + bx_2 + cy_2 + dx_2y_2$$
$$V_3 = a + bx_3 + cy_3 + dx_3y_3$$
$$V_4 = a + bx_4 + cy_4 + dx_4y_4$$

or

$$\begin{bmatrix} V_1 \\ V_2 \\ V_3 \\ V_4 \end{bmatrix} = \begin{bmatrix} 1 & x_1 & y_1 & x_1y_1 \\ 1 & x_2 & y_2 & x_2y_2 \\ 1 & x_3 & y_3 & x_3y_3 \\ 1 & x_4 & y_4 & x_4y_4 \end{bmatrix} \begin{bmatrix} a \\ b \\ c \\ d \end{bmatrix}$$

118

solving for the constants gives

$$V(x,y) = \begin{bmatrix} 1 & x & y & xy \end{bmatrix} \begin{bmatrix} 1 & x_1 & y_1 & x_1y_1 \\ 1 & x_2 & y_2 & x_2y_2 \\ 1 & x_3 & y_3 & x_3y_3 \\ 1 & x_4 & y_4 & x_4y_4 \end{bmatrix}^{-1} \begin{bmatrix} V_1 \\ V_2 \\ V_3 \\ V_4 \end{bmatrix}$$

which is of the form

$$V = \alpha_1 V_1 + \alpha_2 V_2 + \alpha_3 V_3 + \alpha_4 V_4 = \sum_{i=1}^{4} \alpha_i V_i$$

Prob. 6.8 Using the mesh below, the following finite element solution is obtained

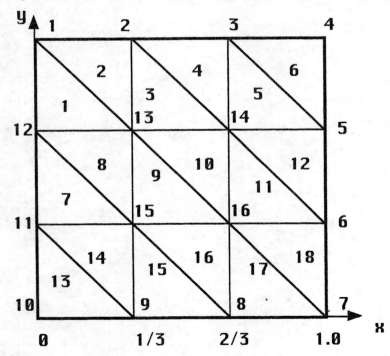

No. of nodes =16, No. of elements = 18, No. of fixed nodes = 12

Node	x	y	Potential
1	0	1.0	0.0
2	0.33	1.0	10.0
3	0.67	1.0	10.0
4	1.0	1.0	0.0
5	1.0	0.67	0.0
6	1.0	0.33	0.0
7	1.0	0.0	0.0
8	0.67	0.0	0.0
9	0.33	0.0	0.0
10	0.0	0.0	0.0
11	0.00	0.33	0.0
12	0.0	0.67	0.0
13	0.33	0.67	3.7502
14	0.67	0.67	3.7502
15	0.33	0.33	1.2498
16	0.67	0.33	1.2498

Prob. 6.9 $V = a + bx + cy$

$$\mathbf{E} = -\nabla V = -(b\mathbf{a}_x + c\mathbf{a}_y)$$

$$\mathbf{E} = -\frac{1}{2A}\left[(y_2 - y_3)V_{e1} + (y_3 - y_1)V_{e2} + (y_1 - y_2)V_{e3}\right]\mathbf{a}_x$$
$$= -\frac{1}{2A}\left[(x_3 - x_2)V_{e1} + (x_1 - x_3)V_{e2} + (x_2 - x_1)V_{e3}\right]\mathbf{a}_y$$

$$\mathbf{E} = -\frac{1}{2A}\sum_{i=1}^{3} P_i V_{ei}\mathbf{a}_x - \frac{1}{2A}\sum_{i=1}^{3} Q_i V_{ei}\mathbf{a}_y$$

$\mathbf{E}$ is uniform within each element and its value is calculated using the formula above.

Prob. 6.10 The results remain almost the same. For the program, input data, and result, see [3] in this chapter.

Prob. 6.11 The exact solution is obtained by using the series expansion technique in section 2.7. Let

$$V_g = \sum_{n=1}^{\infty} A_n \sin \beta x \sinh \beta(c - y)$$

$$V_\ell = \sum_{n=1}^{\infty} \sin \beta x \left[C_n \sinh \beta y - F_p(\cosh \beta y - 1) \right]$$

where g and ℓ denote gas and liquid respectively and $\beta = n\pi/a$.

$$F_p = \begin{cases} \dfrac{4\rho_o a^2}{n^3 \pi^3 \epsilon}, & n = \text{odd} \\ 0, & n = \text{even} \end{cases}$$

$$C_n = F_p \left[\frac{\epsilon_r \sinh \beta b \sinh \beta(c - b) + \cosh \beta(c - b)(\cosh \beta b - 1)}{\epsilon_r \cosh \beta b \sinh \beta(c - b) + \sinh \beta b \cosh \beta(c - b)} \right]$$

$$A_n = C_n \frac{\sinh \beta b}{\sinh \beta(c - b)} - F_p \frac{(\cosh \beta b - 1)}{\sinh \beta(c - b)}$$

For the FEM solution, subroutine GRID is used to generate geometric data. For number of elements $n_e = 100$ and 400, the finite element results are compared with the exact solution as shown below.

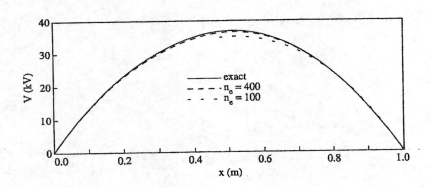

Prob. 6.12 Compare results with those in Example 3.4.

Prob. 6.13 For TE and TM modes, the wave numbers have the exact values given by

$$k_{mn}^2 = (\pi/a)^2 \left[(m + n)^2 + n^2 \right] = \left(\frac{2\pi}{\lambda_c} \right)^2$$

$$\lambda_{c,mn} = \frac{a}{2}\sqrt{(m+n)^2 + n^2}, \ a = 1$$

Prob. 6.14 The exact solution is

$$\lambda_c = \frac{2}{\sqrt{(m/a)^2 + (n/b)^2}}$$

If $a = 2b = 2$cm,

$$\lambda_c = \frac{2a}{\sqrt{m^2 + n^2}} = \frac{4\text{cm}}{\sqrt{m^2 + n^2}}$$

Prob. 6.15 Input data:

NPOIN, NELEM, NNODE, NDIME

18, 3, 3, 2

LNODS, MATNO

1, 2, 3, 10, 16, 15, 14, 9, 1

3, 4, 5, 6, 7, 12, 16, 10, 1

7, 8, 11, 13, 18, 17, 16, 12, 1

JPOIN, COORD

0.0, 0.0

2.0, 0.0

3.0, 0.0

4.0, 0.0

4.15224, 0.76537

4.58579, 1.41421

5.23463, 1.84776

0.0, 2.0

2.0, 2.0

6.0, 2.0

3.55355, 2.64645

6.0, 3.0

0.0, 4.0

```
1.0, 4.0

2.0, 4.0

4.0, 4.0

6.0, 4.0

KBLOC, NDIVX, NDIVY

1

1, 4, 6

WEITX

1, 1, 1, 1

WEITX

1, 1, 1, 1, 1, 1

KBLOC, NDIVX, NDIVY

2

1, 4, 6

WEITX

1, 0.75, 0.5, 0.25

WEITX

1, 1, 1, 1, 1, 1

KBLOC, NDIVX, NDIVY

3

1, 6, 4

WEITX

1, 1, 1, 1, 1, 1

WEITX

0.25, 0.5, 0.75, 1.0
```

With this input data, the region below is divided into finite element mesh.

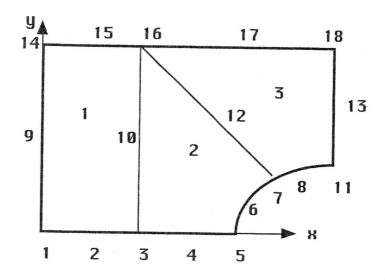

Prob. 6.16 For the original mesh,

$$d_1 = 2, \ d_2 = 3, \ d_3 = 3, \ d_4 = 4, \ d_5 = 4, \ d_6 = 4, \ d_7 = 11, \ d_8 = 3,$$
$$d_9 = 3, \ d_{10} = 4, \ d_{11} = 4, \ d_{12} = 4, \ d_{13} = 4, \ d_{15} = 3, \ d_{16} = 2$$

$B = $ maximum $d_e = 4$.

(b) Renumbering the mesh as shown below gives the minimum bandwidth. In this case,

$$d_1 = 3 = d_2 = \cdots = d_{16}$$

Hence, $B = $ maximum $d_e = 3$.

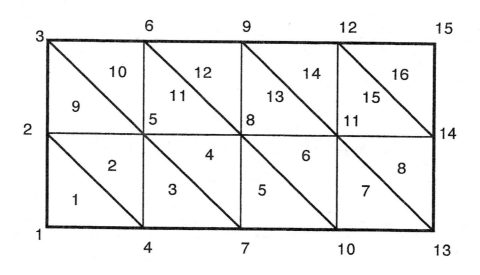

Prob. 6.17 From inspection, $B = $ maximum $d_e = 15 - 1 = 14$ in element 15. With the renumbered mesh shown below, $B = $ maximum $d_e = 4$ in every element.

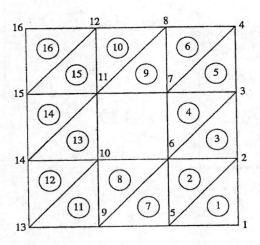

Prob. 6.18 See the solution to Prob. 3.19 and compare the finite element solution with that.

Prob. 6.19 For $n = 2$ in Table 6.7,

$$\alpha_\ell = p_i(\xi_1)p_j(\xi_2)p_k(\xi_3), \qquad \ell = 1, 2, \cdots, 6$$

Since $n = 2$,

$$P_0 = 1, \ P_1 = \xi, \ P_2 = (2\xi - 1)\xi, \ P_3 = \frac{1}{3}(2\xi - 2)(2\xi - 1), \quad \text{etc.}$$

$$\alpha_1 = \alpha_{200} = P_2(\xi_1)P_0(\xi_2)P_0(\xi_3) = \xi_1(2\xi_1 - 1)$$
$$\alpha_4 = \alpha_{020} = \xi_2(2\xi_2 - 1)$$
$$\alpha_6 = \alpha_{002} = \xi_3(2\xi_3 - 1)$$
$$\alpha_2 = \alpha_{110} = (2\xi_1)(2\xi_2).1 = 4\xi_1\xi_2$$
$$\alpha_3 = \alpha_{101} = 4\xi_1\xi_3$$
$$\alpha_5 = \alpha_{011} = 4\xi_2\xi_3$$

Prob. 6.20 (a) $dS = 2dAd\xi_1 d\xi_2$

$$I = \int x\,dS = 2A \int\int (x_1\xi_1 + x_2\xi_2 + x_3\xi_3)d\xi_1 d\xi_2$$

$$= 2A(x_1 + x_2 + x_3) \int\int \xi_1 d\xi_1 d\xi_2$$

$$= 2A(x_1 + x_2 + x_3)\left(\frac{1!0!0!}{3!}\right)$$

$$= A\hat{x}$$

where $\hat{x} = \frac{1}{3}(x_1 + x_2 + x_3)$.

(b)

$$\int_S x^2 dS = \int\int (\xi_1 x_1 + \xi x_2 + \xi_3 x_3)^2 dS$$

$$= \int\int \left(\xi_1^2 x_1^2 + \xi_2^2 x_2^2 + \xi_3^2 x_3^2 + 2\xi_1\xi_2 x_1 x_2 + 2\xi_2\xi_3 x_2 x_3 + 2\xi_1\xi_3 x_1 x_3\right)dS$$

$$= 2A\frac{2}{4!}\left(x_1^2 + x_2^2 + x_3^2 + x_1 x_2 + x_2 x_3 + x_1 x_3\right)$$

$$= \frac{A}{12}\left[x_1^2 + x_2^2 + x_3^2 + (x_1 + x_2 + x_3)^2\right]$$

$$= \frac{A}{12}\left[x_1^2 + x_2^2 + x_3^2 + 9\hat{x}^2\right]$$

(c) Similarly,

$$\int_S xy\,dx\,dy = \frac{A}{12}\left[x_1 y_1 + x_2 y_2 + x_3 y_3 + 9\hat{x}\hat{y}\right]$$

where $\hat{x} = \frac{1}{3}(x_1 + x_2 + x_3)$, $\hat{y} = \frac{1}{3}(y_1 + y_2 + y_3)$.

Prob. 6.21 (a)

$$\int_S \alpha_3^2 dS = 16 \int_S \xi_1^2 \xi_3^2 dS = 32A\frac{(2!)(2!)}{(2+2+0+2)!} = \frac{32A}{180}$$

(b)

$$\int_s \alpha_1 \alpha_5 dS = 4 \int_S \xi_1 \xi_2 \xi_3 (2\xi_1 - 1)dS = 4 \int_S (2\xi_1^2 \xi_2 \xi_3 - \xi_1 \xi_2 \xi_3)dS$$

$$= 8A\left[\frac{2(2!)(1!)(1!)}{(2+1+1+2)!} - \frac{1}{(1+1+1+2)!}\right] = -\frac{A}{15}$$

(c)

$$\int_s \alpha_1 \alpha_2 \alpha_3 dS = 16 \int_S \xi_1^3 \xi_2 \xi_3 (2\xi_1 - 1) dS = 16 \int_S (2\xi_1^4 \xi_2 \xi_3 - \xi_1^3 \xi_2 \xi_3) dS$$

$$= 32A \left[\frac{2(4!)(1!)(1!)}{(4+1+1+2)!} - \frac{3!(1)(1)}{(3+1+1+2)!} \right] = 0$$

Prob. 6.22

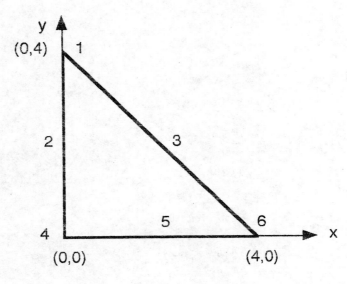

$$A = \frac{1}{2}(4 \times 4) = 8$$

$$\xi_1 = \frac{A_1}{A}, \qquad \xi_2 = \frac{A_2}{A}, \qquad \xi_3 = \frac{A_3}{A}$$

$$A_1 = \begin{vmatrix} 1 & x & y \\ 1 & 0 & 0 \\ 1 & 4 & 0 \end{vmatrix} = 2y$$

$$A_2 = \begin{vmatrix} 1 & x & y \\ 1 & 4 & 0 \\ 1 & 0 & 4 \end{vmatrix} = 8 - 2x - 2y$$

$$A_3 = \begin{vmatrix} 1 & x & y \\ 1 & 0 & 0 \\ 1 & 0 & 4 \end{vmatrix} = 2x$$

$$\xi_1 = \frac{y}{4}, \qquad \xi_2 = \frac{1}{4}(4 - x - y), \qquad \xi_3 = \frac{x}{4}$$

$$\alpha_1 = \xi_1(2\xi_1 - 1) = \frac{y}{8}(y - 2)$$

$$\alpha_2 = 4\xi_1\xi_2 = \frac{1}{4}y(4 - x - y)$$

$$\alpha_3 = 4\xi_1\xi_3 = \frac{1}{4}xy$$

$$\alpha_4 = \xi_2(2\xi_2 - 1) = \frac{1}{8}(4 - x - y)(2 - x - y)$$

$$\alpha_5 = 4\xi_2\xi_3 = \frac{x}{4}y(4 - x - y)$$

$$\alpha_6 = \xi_3(2\xi_3 - 1) = \frac{x}{8}(x - 2)$$

Prob. 6.23

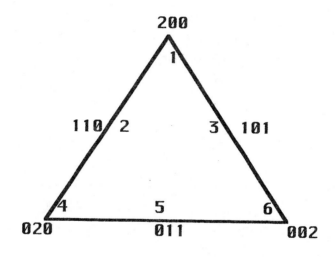

$$T_{11} = T_{200,200} = T_{002,002} = T_{020,020} = T_{44} = T_{66}$$

$$T_{12} = T_{13} = T_{42} = T_{45} = T_{63} = T_{65}$$

$$T_{22} = T_{110,110} = T_{33} = T_{55}$$

$$T_{14} = T_{10} = T_{41} = T_{46} = T_{61} = T_{64} = T_{15} = T_{43} = T_{62}$$

$$T_{24} = T_{36} = T_{21} = T_{56} = T_{54}$$

$$T_{26} = T_{34} = T_{26} = T_{34} = T_{51}$$

$$T_{46} = T_{64} = T_{16} = T_{14} = T_{61} = T_{41}$$

$$T_{11} = \int\int \alpha_1\alpha_2 dS = 2A \int_0^1 \int_1^{1-\xi_2} \xi_1^2(2\xi_1 - 1)^2 d\xi_1 d\xi_2$$

$$= 2A \int_0^1 \left(\frac{4\xi_1^5}{5} - \frac{4\xi_1^4}{4} + \frac{\xi_1^2}{3}\right)\Big|_0^{1-\xi_2} d\xi_2 = \frac{6A}{180}$$

$$T_{12} = 2A \int_0^1 \int_1^{1-\xi_2} \xi_1(2\xi_1 - 1)(4\xi_1\xi_2)d\xi_1 d\xi_2 = 2A \cdot 4 \int_0^1 \left(\frac{2\xi_1^4}{4} - \frac{\xi_1^3}{3}\right)\bigg|_0^{1-\xi_2} d\xi_2 = 0$$

$$T_{14} = 2A \int_0^1 \int_1^{1-\xi_2} \xi_1(2\xi_1 - 1)\xi_2(2\xi_2 - 1)d\xi_1 d\xi_2 = -\frac{A}{180}$$

$$T_{22} = 2A \int_0^1 \int_1^{1-\xi_2} (4\xi_1\xi_2)^2 d\xi_1 d\xi_2 = \frac{32A}{180}$$

$$T_{23} = 2A \int_0^1 \int_1^{1-\xi_2} (4\xi_1\xi_2)(4\xi_1\xi_3)d\xi_1 d\xi_2 = \frac{16A}{180}$$

Hence,

$$T = \frac{A}{180}\begin{bmatrix} 6 & 0 & 0 & -1 & 4 & -1 \\ 0 & 32 & 16 & 0 & 16 & -4 \\ 0 & 16 & 32 & -4 & 16 & 0 \\ -1 & 0 & -4 & 6 & 0 & -1 \\ -4 & 16 & 16 & 0 & 32 & 0 \\ -1 & -4 & 0 & -1 & 0 & 6 \end{bmatrix}$$

Prob. 6.24 (a) For $n = 1$, $\alpha_1 = \xi_1$, $\alpha_2 = \xi_2$, $\alpha_3 = \xi_3$.

$$Q_{11}^{(2)} = \int_0^1 \int_0^{1-\xi_1} \left(\frac{\partial \alpha_1}{\partial \xi_3} - \frac{\partial \alpha_1}{\partial \xi_1}\right)\left(\frac{\partial \alpha_1}{\partial \xi_3} - \frac{\partial \alpha_1}{\partial \xi_1}\right)d\xi_1 d\xi_2$$

$$= \int_0^1 \int_0^{1-\xi_1} (-1)(-1)d\xi_1 d\xi_2 = \frac{1}{2}$$

$$Q_{22}^{(2)} = 0 = Q_{12}^{(2)} = Q_{23}^{(2)}$$

$$Q_{33}^{(2)} = \int_0^1 \int_0^{1-\xi_1} \left(\frac{\partial \alpha_3}{\partial \xi_2} - \frac{\partial \alpha_3}{\partial \xi_1}\right)\left(\frac{\partial \alpha_3}{\partial \xi_3} - \frac{\partial \alpha_3}{\partial \xi_1}\right)d\xi_1 d\xi_2$$

$$= \int_0^1 \int_0^{1-\xi_1} (1)d\xi_1 d\xi_2 = \frac{1}{2}$$

$$Q_{13}^{(2)} = \int_0^1 \int_0^{1-\xi_1} (1)d\xi_1 d\xi_2 = -\frac{1}{2}$$

Since $Q_{ij}^{(q)} = Q_{ji}^{(q)}$,

$$Q^{(2)} = \frac{1}{2}\begin{bmatrix} 1 & 0 & -1 \\ 0 & 0 & 0 \\ -1 & 0 & 1 \end{bmatrix}$$

$$Q^{(3)} = RQ^{(2)}R = \frac{R}{2}\begin{bmatrix} 1 & 0 & -1 \\ 0 & 0 & 0 \\ -1 & 0 & 1 \end{bmatrix}\begin{bmatrix} 0 & 0 & 1 \\ 1 & 0 & 0 \\ 0 & 1 & 0 \end{bmatrix}$$

$$= \frac{1}{2}\begin{bmatrix} 0 & 0 & 1 \\ 1 & 0 & 0 \\ 0 & 1 & 0 \end{bmatrix}\begin{bmatrix} 0 & -1 & 1 \\ 0 & 0 & 0 \\ 0 & 1 & -1 \end{bmatrix}$$

$$= \frac{1}{2}\begin{bmatrix} 0 & 1 & -1 \\ 0 & -1 & 1 \\ 0 & 0 & 0 \end{bmatrix}$$

(b) For $n = 2$,

$$\alpha_1 = \xi_1(2\xi_2 - 1), \ \alpha_2 = 4\xi_1\xi_3, \ \alpha_3 = 4\xi_1\xi_3,$$
$$\alpha_4 = \xi_2(2\xi_2 - 1), \ \alpha_5 = 4\xi_2\xi_3, \ \alpha_6 = \xi_3(2\xi_2 - 1)$$

$$Q_{11}^{(2)} = \int_0^1 \int_0^{1-\xi_2} (-4\xi_1 + 1)^2 d\xi_1 d\xi_2 = \frac{3}{6}$$

$$Q_{22}^{(2)} = \int_0^1 \int_0^{1-\xi_2} (-4\xi_1 + 1)^2 d\xi_1 d\xi_2 = \frac{8}{6}$$

$$Q_{33}^{(2)} = \int_0^1 \int_0^{1-\xi_2} (4\xi_1 - 4\xi_3)^2 d\xi_1 d\xi_2 = \frac{8}{6}$$

$$Q_{44}^{(2)} = 0$$

$$Q_{55}^{(2)} = \int_0^1 \int_0^{1-\xi_2} (4\xi_2)^2 d\xi_1 d\xi_2 = \frac{8}{6}$$

$$Q_{66}^{(2)} = \int_0^1 \int_0^{1-\xi_2} (4\xi_2 - 1)^2 d\xi_1 d\xi_2 = \frac{3}{6}$$

$$Q_{12}^{(2)} = \int_0^1 \int_0^{1-\xi_2} (-4\xi_2 - 1)(-4\xi_2) d\xi_1 d\xi_2 = 0$$

$$Q_{13}^{(2)} = \int_0^1 \int_0^{1-\xi_2} (-4\xi_1 - 1)(4\xi_1 - 4\xi_2) d\xi_1 d\xi_2 = \frac{4}{6}$$

Since $Q_{ij}^{(q)} = Q_{ji}^{(q)}$,

$$Q^{(2)} = \frac{1}{6}\begin{bmatrix} 3 & 0 & -4 & 0 & 0 & 1 \\ 0 & 8 & 0 & 0 & -8 & 0 \\ -4 & 0 & 8 & 0 & 0 & -4 \\ 0 & 0 & 0 & 0 & 0 & 0 \\ 0 & -8 & 0 & 0 & 8 & 0 \\ 1 & 0 & -4 & 0 & 0 & 3 \end{bmatrix}$$

$$Q^{(3)} = Q^{(3)} = RQ^{(2)}R = \frac{1}{6}\begin{bmatrix} 3 & -4 & 0 & 1 & 0 & 0 \\ -4 & 8 & 0 & -4 & 0 & 0 \\ 0 & 0 & 8 & 0 & -8 & 0 \\ 1 & -4 & 0 & 3 & 0 & 0 \\ 0 & 0 & -8 & 0 & 8 & 0 \\ 0 & 0 & 0 & 0 & 0 & 0 \end{bmatrix}$$

Prob. 6.25 For $n = 1$, the triangle is as shown in (a).

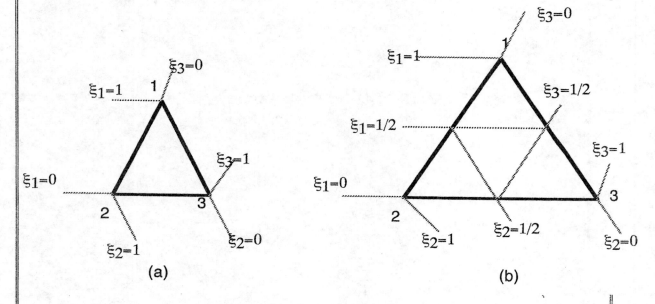

(a)   (b)

$$\alpha_1 = \xi_1, \qquad \alpha_2 = \xi_2, \qquad \alpha_3 = \xi_3$$

$$D_{ij}^{(1)} = \left.\frac{\partial \alpha_1}{\partial \xi_1}\right|_{P_j}$$

Hence,

$$D_{ij}^{(1)} = 1, \qquad D_{2j}^{(1)} = \frac{\partial \alpha_2}{\partial \xi_1} = 0 = D_{3j}^{(1)}$$

and

$$D^{(1)} = \begin{bmatrix} 1 & 1 & 1 \\ 0 & 0 & 0 \\ 0 & 0 & 0 \end{bmatrix}$$

$$D^{(2)} = RD^{(1)}R = \begin{bmatrix} 0 & 0 & 0 \\ 1 & 1 & 1 \\ 0 & 0 & 0 \end{bmatrix}$$

$$D^{(3)} = RD^{(2)}R = \begin{bmatrix} 0 & 0 & 0 \\ 0 & 0 & 0 \\ 1 & 1 & 1 \end{bmatrix}$$

For $n = 2$, the triangle is as shown in (b).

$$D_{1j}^{(1)} = \frac{\partial \alpha_1}{\partial \xi_1}\bigg|_{P_j} = 4\xi_1 - 1\bigg|_{P_j}$$

$$D_{11}^{(1)} = 3, \qquad D_{12}^{(1)} = D_{13}^{(1)} = 1, \quad D_{14}^{(1)} = D_{15}^{(1)} = D_{16}^{(1)} = -1$$

$$D_{2j}^{(1)} = \frac{\partial \alpha_2}{\partial \xi_1}\bigg|_{P_j} = 4\xi_2\bigg|_{P_j}$$

$$D_{21}^{(1)} = D_{23}^{(1)} = D_{26}^{(1)} = 0, \quad D_{22}^{(1)} = D_{25}^{(1)} = 2, \qquad D_{24}^{(1)} = 4$$

$$D_{3j}^{(1)} = \frac{\partial \alpha_3}{\partial \xi_1}\bigg|_{P_j} = 4\xi_3\bigg|_{P_j}$$

$$D_{31}^{(1)} = D_{32}^{(1)} = D_{34}^{(1)} = 0, \quad D_{33}^{(1)} = D_{35}^{(1)} = 2, \qquad D_{36}^{(1)} = 4$$

$$D_{4j}^{(1)} = \frac{\partial \alpha_4}{\partial \xi_1}\bigg|_{P_j} = 0 = D_{5j}^{(1)} = D_{6j}^{(1)}$$

Thus

$$D^{(1)} = \begin{bmatrix} 3 & 1 & 1 & -1 & -1 & -1 \\ 0 & 2 & 0 & 4 & 2 & 0 \\ 0 & 0 & 2 & 0 & 2 & 4 \\ 0 & 0 & 0 & 0 & 0 & 0 \\ 0 & 0 & 0 & 0 & 0 & 0 \\ 0 & 0 & 0 & 0 & 0 & 0 \end{bmatrix}$$

$$D^{(2)} = RD^{(1)}R = \begin{bmatrix} 0 & 0 & 0 & 0 & 0 & 0 \\ 4 & 2 & 0 & 0 & 0 & 0 \\ 0 & 0 & 0 & 0 & 0 & 0 \\ -1 & 1 & -1 & 3 & 1 & -1 \\ 0 & 0 & 2 & 0 & 2 & 4 \\ 0 & 0 & 0 & 0 & 0 & 0 \end{bmatrix}$$

$$D^{(3)} = RD^{(2)}R = \begin{bmatrix} 0 & 0 & 0 & 0 & 0 & 0 \\ 0 & 0 & 0 & 0 & 0 & 0 \\ 4 & 2 & 0 & 0 & 0 & \\ 0 & 0 & 0 & 0 & 0 & 0 \\ 0 & 2 & 0 & 4 & 2 & 0 \\ -1 & -1 & 1 & -1 & 1 & 3 \end{bmatrix}$$

Prob. 6.26 (a) If

$$D_{1j}^{(q)} = \frac{\partial \alpha_1}{\partial \xi_q}\bigg|_{P_j}$$

then

$$\frac{\partial \alpha_1}{\partial \xi_q} = \sum_{j=1}^{m} D_{1j}^{(q)} \alpha_j$$

Hence,

$$K_{ij}^{(q)} = \int \int \frac{\partial \alpha_i}{\partial \xi_q} \frac{\partial \alpha_j}{\partial \xi_q} \, dS$$

$$= \int \int \sum_{k=1}^{m} \sum_{l=1}^{m} D_{ik}^{(p)} \alpha_k D_{jl}^{(q)} \alpha_l \, dS$$

But

$$\int \int \alpha_i \alpha_j \, dS = T_{ij}$$

Thus

$$K_{ij}^{(q)} = \sum_{k=1}^{m} \sum_{l=1}^{m} D_{ik}^{(p)} T_{kl} D_{jl}^{(q)}$$

In matrix form,

$$K(pq) = D^{(p)} T D^{(q)t} \tag{1}$$

(b)

$$Q_{ij}^{(q)} = \int \int \left(\frac{\partial \alpha_i}{\partial \xi_{q+1}} - \frac{\partial \alpha_j}{\partial \xi_{q-1}} \right) \left(\frac{\partial \alpha_j}{\partial \xi_{q+1}} - \frac{\partial \alpha_j}{\partial \xi_{q-1}} \right) d\xi_1 d\xi_2$$

Since $d\xi_1 d\xi_2 = \dfrac{dS}{2A}$

$$Q_{ij}^{(q)} = \frac{1}{2A} \int \int \left(\frac{\partial \alpha_i}{\partial \xi_{q+1}} - \frac{\partial \alpha_j}{\partial \xi_{q-1}} \right) \left(\frac{\partial \alpha_j}{\partial \xi_{q+1}} - \frac{\partial \alpha_j}{\partial \xi_{q-1}} \right) dS \tag{2}$$

Introducing (2) in (1) gives

$$Q(q) = \frac{1}{2A} \left(D^{(q+1)} - D^{(q-1)} \right) T \left(D^{(q+1)} - D^{(q-1)} \right)^t$$

Hence,

$$Q(1) = \frac{1}{2A} \left(D^{(2)} - D^{(3)} \right) T \left(D^{(2)} - D^{(3)} \right)^t \tag{3}$$

For $n = 1$,

$$Q(1) = \frac{1}{2A} \left(D^{(2)} - D^{(3)} \right) T \left(\begin{bmatrix} 0 & 0 & 0 \\ 1 & 1 & 1 \\ 0 & 0 & 0 \end{bmatrix} - \begin{bmatrix} 0 & 0 & 0 \\ 0 & 0 & 0 \\ 1 & 1 & 1 \end{bmatrix} \right)^t$$

$$= \frac{1}{2A} \left(D^{(2)} - D^{(3)} \right) \frac{A}{12} \begin{bmatrix} 2 & 1 & 1 \\ 1 & 2 & 1 \\ 1 & 1 & 2 \end{bmatrix} \begin{bmatrix} 0 & 1 & -1 \\ 0 & 1 & -1 \\ 0 & 1 & -1 \end{bmatrix}$$

$$= \frac{1}{24} \begin{bmatrix} 0 & 0 & 0 \\ 1 & 1 & 1 \\ -1 & -1 & -1 \end{bmatrix} \begin{bmatrix} 0 & 4 & -4 \\ 0 & 4 & -4 \\ 0 & 4 & -4 \end{bmatrix}$$

$$Q(1) = \frac{1}{2} \begin{bmatrix} 0 & 0 & 0 \\ 0 & 1 & -1 \\ 0 & -1 & -1 \end{bmatrix}$$

133

For $n = 2$,

$$T = \frac{A}{180}\begin{bmatrix} 6 & 0 & 0 & -1 & -4 & -1 \\ 0 & 32 & 16 & 0 & 16 & -4 \\ 0 & 16 & 32 & -4 & 16 & 0 \\ -1 & 0 & -4 & 6 & 0 & -1 \\ -4 & 16 & 16 & 0 & 32 & 0 \\ -1 & -4 & 0 & -1 & 0 & 6 \end{bmatrix} \tag{4}$$

while $D^{(2)}$ and $D^{(3)}$ are given in the previous problem. Substituting $D^{(2)}$ and $D^{(3)}$ along with (4) into (3) yields

$$Q(1) = \frac{1}{6}\begin{bmatrix} 0 & 0 & 0 & 0 & 0 & 0 \\ 0 & 8 & -8 & 0 & 0 & 0 \\ 0 & -8 & 8 & 0 & 0 & 0 \\ 0 & 0 & 0 & 3 & -4 & 1 \\ 0 & 0 & 0 & -4 & 8 & -4 \\ 0 & 0 & 0 & 1 & -4 & 3 \end{bmatrix}$$

Prob. 6.27

$$\alpha_q = \alpha_{ijkl} = P_i(\xi_1)P_j(\xi_2)P_k(\xi_3)P_l(\xi_4)$$
$$\alpha_1 = \alpha_{2000} = \xi_1(2\xi_2 - 1)$$
$$\alpha_5 = \alpha_{0200} = \xi_2(2\xi_2 - 1)$$
$$\alpha_8 = \alpha_{0020} = \xi_3(2\xi_3 - 1)$$
$$\alpha_{10} = \alpha_{0002} = \xi_4(2\xi_4 - 1)$$
$$\alpha_2 = \alpha_{1100} = (2\xi_1)(2\xi_2)(1)(1) = 4\xi_1\xi_2$$
$$\alpha_3 = \alpha_{1010} = 4\xi_1\xi_2$$
$$\alpha_4 = \alpha_{1000} = 4\xi_1\xi_4$$
$$\alpha_6 = \alpha_{0110} = 4\xi_2\xi_3$$
$$\alpha_7 = \alpha_{0101} = 4\xi_2\xi_4$$
$$\alpha_9 = \alpha_{0011} = 4\xi_3\xi_4$$

Prob. 6.28

$$\int_v z^2 dv = \int_v \left(\xi_1 z_1 + \xi_2 z_2 + \xi_3 z_3 + \xi_4 z_4 \right)^2 dv$$

$$= \int_v \left(\xi_1^2 z_1^2 + \xi_2^2 z_2^2 + \xi_4^4 z^4 + 2\xi_1 \xi_2 z_1 z_2 + 2\xi_1 \xi_3 z_1 z_3 \right.$$

$$\left. + 2\xi_1 \xi_4 z_1 z_4 + 2\xi_2 \xi_3 z_2 z_3 + 2\xi_2 \xi_4 z_3 z_4 \right)^2 dv$$

$$= \frac{2}{5!} (6v) \left(z_1^2 + z_2^2 + z_3^2 + z_4^2 + z_1 z_2 + z_1 z_3 + z_1 z_4 + z_2 z_3 + z_2 z_4 + z_3 z_4 \right)$$

$$= \frac{v}{10} \left(z_1^2 + z_2^2 + z_3^2 + z_4^2 + z_1 z_2 + z_1 z_3 + z_1 z_4 + z_2 z_3 + z_2 z_4 + z_3 z_4 \right)$$

If the center is at the centroid,

$$z_1 + z_2 + z_3 + z_4 = 0$$

and hence

$$\int_v z^2 dv = \frac{v}{20} \left(z_1^2 + z_2^2 + z_3^2 + z_4^2 \right)$$

Prob. 6.29 By definition,

$$T_{ij} = \int_v \alpha_i \alpha_j \; dv$$

For the first-order element,

$$\alpha_1 = \xi_1, \;\; \alpha_2 = \xi_2, \;\; \alpha_3 = \xi_3, \;\; \alpha_4 = \xi_4$$

$$T_{11} = \int_v \xi_1^2 \; dv = \frac{2!}{(2+3)!} 6v = \frac{v}{10}$$

$$T_{12} = \int_v \xi_1 \xi_2 \; dv = \frac{1!}{(1+1+3)!} 6v = \frac{v}{20}$$

In general,

$$T_{ij} = \begin{cases} \dfrac{v}{20}, & i \neq j \\[2mm] \dfrac{v}{10}, & i = j \end{cases}$$

Thus,

$$T = \frac{v}{20} \begin{bmatrix} 2 & 1 & 1 & 1 \\ 1 & 2 & 1 & 1 \\ 1 & 1 & 2 & 1 \\ 1 & 1 & 1 & 2 \end{bmatrix}$$

Prob. 6.30

$$M_{11} = \frac{1}{v} \int_v \xi_1^2 \; dv = \frac{1}{v} \frac{2!0!0!0!}{(2+0+0+0+3)!} 6v = \frac{2}{20}$$

$$M_{12} = \frac{1}{v} \int_v \xi_1 \xi_2 \, dv = \frac{1}{v} \frac{1!1!0!0!}{(1+1+0+0+3)!} 6v = \frac{1}{20}$$

Thus in general,

$$M_{ij} = \begin{cases} \dfrac{2}{20}, & i \neq j \\[2mm] \dfrac{1}{20}, & i = j \end{cases}$$

i.e.

$$M = \frac{1}{20} \begin{bmatrix} 2 & 1 & 1 & 1 \\ 1 & 2 & 1 & 1 \\ 1 & 1 & 2 & 1 \\ 1 & 1 & 1 & 2 \end{bmatrix}$$

Prob. 6.31

$$B_1 = \frac{\partial}{\partial \rho} + jk + \frac{1}{2\rho}$$

$$B_2 = \left(\frac{\partial}{\partial \rho} + jk + \frac{5}{2\rho} \right) \left(\frac{\partial}{\partial \rho} + jk + \frac{1}{2\rho} \right)$$

This produces a second-order derivative, which is conveniently eliminated by the substitution of the Helmholtz's equation

$$\frac{\partial^2}{\partial r^2} = -k^2 - \frac{1}{r} \frac{\partial}{\partial r} - \frac{1}{r^2} \frac{\partial^2}{\partial \phi^2}$$

The result is

$$B_2 = \frac{\partial}{\partial \rho} + jk + \frac{1}{2\rho} - \frac{1}{8\rho(1 + jk\rho)} - \frac{1}{2\rho(1 + jk\rho)} \frac{\partial^2}{\partial \phi^2}$$

CHAPTER 7

Prob. 7.1

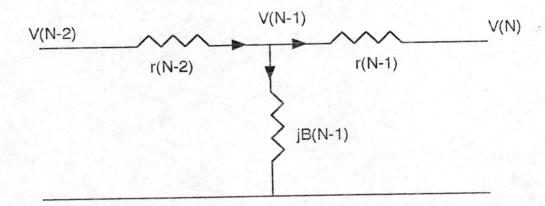

Consider the network shown. Applying Kirchhoff's current law to the node,

$$\frac{V(N-1) - V(N-2)}{r(N-1)} = \frac{V(N) - V(N-1)}{r(N-1)} + jB(N-1)(0 - V(N-1))$$

$$V(N) = V(N-1)\frac{r(N-1)}{r(N-2)}[V(N-1) - V(N-2)] + jB(N-1)V(N-1)r(N-1)$$

Prob. 7.2

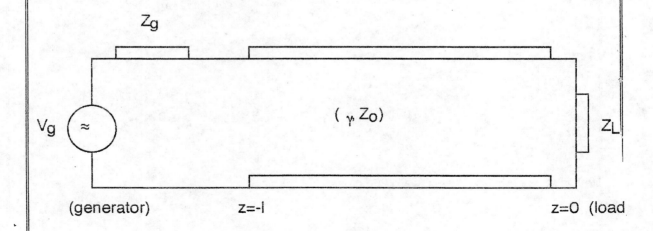

Consider the lossy line shown. At any point on the line,

$$V_s(z) = V_o^+(e^{-\gamma z} + \Gamma_L e^{\gamma z})$$

$$I_s(z) = \frac{V_o^+}{Z_o}(e^{-\gamma z} - \Gamma_L e^{\gamma z})$$

We now apply these to Fig. 7.30.

$$V_1 = V_s(z = -l) = V_o^+(e^{\gamma l} + \Gamma_L e^{-\gamma l}) \tag{1}$$

$$V_2 = V_s(z = 0) = V_o^+(1 + \Gamma_L) \tag{2}$$

$$I_1 = I_s(z = -l) = \frac{V_o^+}{Z_o}(e^{\gamma l} - \Gamma_L e^{-\gamma l}) \tag{3}$$

$$I_2 = -I_s(z = 0) = -\frac{V_o^+}{Z_o}(1 - \Gamma_L) \tag{4}$$

From (1) to (4),

$$\begin{aligned}
V_1 &= V_o^+(1 + \Gamma_L)\cosh\gamma l + V_o^+(1 - \Gamma_L)\sinh\gamma l \\
&= V_2\cosh\gamma l - I_2 Z_o \sinh\gamma l
\end{aligned} \tag{5}$$

$$\begin{aligned}
I_1 &= \frac{V_o^+}{Z_o}(1 - \Gamma_L)\cosh\gamma l + \frac{V_o^+}{Z_o}(1 + \Gamma_L)\sinh\gamma l \\
&= -I_2\cosh\gamma l + \frac{V_2}{Z_o}\sinh\gamma l
\end{aligned} \tag{6}$$

Putting (5) and (6) in matrix form,

$$\begin{bmatrix} V_1 \\ I_1 \end{bmatrix} = \begin{bmatrix} \cosh\gamma l & Z_o\sinh\gamma l \\ \frac{1}{Z_o}\sinh\gamma l & \cosh\gamma l \end{bmatrix} \begin{bmatrix} V_2 \\ -I_2 \end{bmatrix}$$

as required.

Prob. 7.3 The equations governing the voltage and current in the transmission line section are

$$-\frac{\partial V}{\partial z}\frac{\Delta z}{2} = IR\Delta z/2 + L\Delta z/2\frac{\partial I}{\partial t} \tag{2}$$

$$\frac{\partial I}{\partial z}\Delta z = I_m - C\Delta z\frac{\partial V}{\partial t} \tag{2}$$

Rearranging gives

$$-\frac{\partial V}{\partial z} = IR + L\frac{\partial I}{\partial t} \tag{3}$$

$$\frac{\partial I}{\partial z} = \frac{I_m}{\Delta z} - C\frac{\partial V}{\partial t} \tag{4}$$

Differentiating (3) with z and using (4) gives

$$\frac{\partial^2 V}{\partial z^2} = -Ri + RC\frac{\partial V}{\partial t} - L\frac{\partial i}{\partial t} + LC\frac{\partial^2 V}{\partial t^2}$$

Prob. 7.4

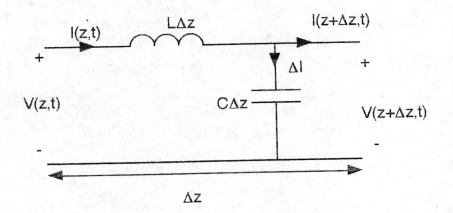

Consider the model shown.

$$I(z,t) = I(z + \Delta z, t) + \Delta I = I(z + \Delta z, t) + C\Delta z \frac{dV(z + \Delta z, t)}{dt}$$

$$-\frac{I(z + \Delta z, t) - I(z,t)}{\Delta z} = C\frac{dV(z + \Delta z, t)}{dt}$$

As $\Delta \rightarrow 0$,

$$-\frac{\partial I}{\partial z} = C\frac{\partial V}{\partial t} \tag{1}$$

Similarly,

$$-\frac{\partial V}{\partial z} = L\frac{\partial V}{\partial t} \tag{2}$$

From (1) and (2),

$$\frac{\partial^2 \Phi}{\partial z^2} = LC\frac{\partial^2 \Phi}{\partial t^2} \tag{3}$$

where Φ is either I or V.

For TEM modes, Maxwell's equations become ($E_z = E_y = 0$, $H_x = H_z = 0$)

$$\nabla \times \mathbf{E} = -\mu\frac{\mathbf{H}}{\partial t} \qquad \rightarrow \qquad -\frac{\partial E_x}{\partial z} = \mu\frac{\partial H_y}{\partial t} \tag{4}$$

$$\nabla \times \mathbf{H} = \mathbf{J} + \epsilon\frac{\mathbf{E}}{\partial t} \qquad \rightarrow \qquad -\frac{\partial H_y}{\partial z} = \epsilon\frac{\partial E_x}{\partial t} \tag{5}$$

Combining (4) and (5)

$$\frac{\partial^2 \Phi}{\partial z^2} = \mu\epsilon \frac{\partial^2 \Phi}{\partial x^2} \qquad (6)$$

Comparing (1) to (3) with (4) to (6), we obtain the following equivalencies:

$$V \equiv E_x$$
$$I \equiv H_y$$
$$C \equiv \epsilon$$
$$L \equiv \mu$$

Prob. 7.5 Modify the program by inserting the following lines:

C ** INSERT THE FOLLOWING BOUNDARY CONDITIONS

```
    IF(J.EQ.5) VI(IT,I,5,3) = - VR(IT,I,5,3)

    IF(J.EQ.1) VI(IT,1,1,1) = VR(IT,1,1,1)

    IF(I.EQ.1) VI(IT,1,I,2) = VR(IT,I,J,2)

    IF(I.EQ.5) VI(IT,5,J,4) = - VR(IT,5,J,4)
```

With this modification, the TLM solution is

$$\Delta/\lambda = 0.0501$$

compared with the exact solution of $\Delta/\lambda = 0.05$.

Prob. 7.6 For this case, the program in Fig. 7.14 is modified as follows.

C ** INSERT THE FOLLOWING BOUNDARY CONDITIONS

```
    IF(J.EQ.10) VI(IT,I,10,3) = - VR(IT,I,10,3)

    IF(J.EQ.1) VI(IT,1,1,1) =- VR(IT,1,1,1)

    IF(I.EQ.1) VI(IT,1,J,2) = -VR(IT,I,J,2)

    IF(I.EQ.5) VI(IT,10,J,4) = - VR(IT,10,J,4)
```

The results are shown below.

Mode	TLM	Analytic
TM_{12}	0.0788	0.0791
TM_{22}	0.0999	0.1
TM_{13}	0.1106	0.1118
TM_{23}	0.1269	0.1275
TM_{14}	0.1422	0.1458
$\vdots$	$\vdots$	$\vdots$
TM_{44}	0.1995	0.2

Prob. 7.7 The plot [see Fig. 9 of Ref. 9] is sketched below.

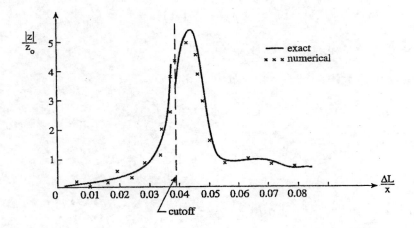

The exact solution is for infinitely long guide. In the evanescent mode,

$$\frac{|Z|}{Z_o} = \frac{1}{\sqrt{(\lambda_n/\lambda_{nc})^2 - 1}}$$

while in the propagation mode,

$$\frac{|Z|}{Z_o} = \frac{\sqrt{2}(1 + \tan^2 \beta x)}{1 + 2\tan^2 \beta x}, \qquad \beta = \frac{2\pi}{\lambda}$$

Prob. 7.8 Run the program in Fig. 7.19 and take the output at $x = 6$, $z = 13$.

Prob. 7.10 The plot [see Fig. 2 of Ref. 23] is sketched below.

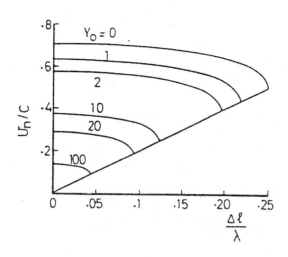

Prob. 7.11

$$\Gamma = \frac{Z_c - Z_o}{Z_c + Z_o} = -1 + \frac{2Z_c}{Z_o + Z_c}$$

$$\simeq -1 + 2\frac{Z_c}{Z_o}, \qquad \text{if } Z_c << Z_o$$

$$= -1 + 2\sqrt{\frac{\mu\omega}{2\sigma_c}}(1+j)\sqrt{\frac{\epsilon_o}{\mu_o}}$$

If $\mu = \mu_o$ and the imaginary part of Γ is negligible,

$$\Gamma \simeq -1 + 2\sqrt{\frac{\epsilon_o\omega}{2\sigma_c}} = -1 + \sqrt{\frac{2\epsilon\omega}{\sigma_c}}$$

Prob. 7.12

$$T \cdot T = \begin{bmatrix} \cos^2\theta/4 - \sin^2\theta/4 & 2j\cos\theta/4\sin\theta/4 \\ 2j\cos\theta/4\sin\theta/4 & \cos^2\theta/4 - \sin^2\theta/4 \end{bmatrix} = \begin{bmatrix} \cos\theta/2 & j\sin\theta/2 \\ j\sin\theta/2 & j\sin\theta/2 \end{bmatrix}$$

$$\begin{bmatrix} 1 & 0 \\ X & 1 \end{bmatrix} \cdot T = \begin{bmatrix} 1 & 0 \\ X & 1 \end{bmatrix} \begin{bmatrix} \cos\theta/4 & j\sin\theta/4 \\ j\sin\theta/4 & j\sin\theta/4 \end{bmatrix} \cdot$$

$$\begin{bmatrix} \cos\theta/4 & j\sin\theta/4 \\ X\cos\theta/4 + j\sin\theta/4 & j\sin\theta/4 + \cos\theta/4 \end{bmatrix}$$

where

$$X = g_o + j(2 + y_o) \tan \theta/2, \qquad Y = j(2 + Z_o) \tan \theta/2$$

$$T \cdot \begin{bmatrix} 1 & Y \\ 0 & 1 \end{bmatrix} T \cdot T$$

$$= T \cdot \begin{bmatrix} 1 & Y \\ 0 & 1 \end{bmatrix} \begin{bmatrix} \cos \theta/2 & j \sin \theta/2 \\ j \sin \theta/2 & j \sin \theta/2 \end{bmatrix}$$

$$= T \cdot \begin{bmatrix} \cos \theta/2 + jY \sin \theta/2 & j \sin \theta/2 + Y \cos \theta/2 \\ j \sin \theta/2 & \cos \theta/2 \end{bmatrix}$$

$$\begin{bmatrix} V_i \\ I_i \end{bmatrix} = T \cdot \begin{bmatrix} \cos \theta/2 + jY \sin \theta/2 & j \sin \theta/2 + Y \cos \theta/2 \\ j \sin \theta/2 & \cos \theta/2 \end{bmatrix} \cdot$$

$$\begin{bmatrix} \cos \theta/4 & j \sin \theta/4 \\ X \cos \theta/4 + j \sin \theta/4 & j \sin \ theta/4 + \cos \theta/4 \end{bmatrix} \begin{bmatrix} V_{i+1} \\ I_{i+1} \end{bmatrix}$$

Prob. 7.13

$$Z_x = \frac{2\Delta l}{u_o \Delta t} - 4 = \frac{0.2}{3 \times 10^8 \Delta} - 4$$

For stability,

$$Z_x = \frac{2\Delta l}{u_o \Delta t} \geq 4$$

or

$$\Delta t \leq \frac{\Delta l}{2u_o} = \frac{0.1}{2 \times 3 \times 10^8} = \frac{1}{6} \text{ ns}$$

Prob. 7.14 The results [26] are tabulated below.

$\delta l/\lambda$	$|Z|$		arg (Z)	
	TLM	exact	TLM	exact
0.023	0.2296	482.04	-0.9283	-1.5708
0.025	2.2531	2.6599	1.8694	1.5708
0.027	0.2550	0.2527	-1.5309	-1.5708
0.029	6.8085	6.6448	1.5923	1.5708
0.031	0.2706	0.2826	-1.5582	-1.5708
0.033	1.8526	1.8320	1.5787	1.5708
0.035	0.4575	0.8521	-1.5569	-1.5708
0.037	0.4875	0.4796	1.5548	1.5708
0.039	5.9766	6.0828	-1.5633	-1.5708
0.041	0.2076	0.2118	-1.5488	1.5708

143

Prob. 7.15 The results [26] are shown below.

$\delta l/\lambda$	$|Z|$		$\arg(Z)$	
	TLM	exact	TLM	exact
0.023	4.1961	6.1272	-0.2806	-0.0106
0.025	2.3822	2.4898	1.2546	1.0610
0.027	0.3281	0.3252	-0.7952	-0.8554
0.029	5.2724	5.1637	0.8459	0.8678
0.031	0.2963	0.3039	-1.1340	-1.1610
0.033	1.8117	1.8038	1.3408	1.3384
0.035	0.8505	0.8529	-1.3820	-1.4025
0.037	0.4912	0.4838	1.3914	1.3932
0.039	5.3772	5.4883	-1.1022	-1.1155
0.041	0.2115	0.2179	-1.2795	-1.3174

Prob. 7.16 The solution [23] is shown below.

Mode	TLM	exact
TE_{10}	0.03721	0.03720
TM_{11}	0.03838	0.03850

Prob. 7.17 The TLM solution [23] is:

for $\epsilon_r = 2$, $k_c a = 1.303$

for $\epsilon_r = 8$, $k_c a = 0.968$

Prob. 7.18 Modify the program in Fig. 7.26 as shown in [22, p. 72] and obtain

$$k_c a = 5.5292$$

144

Prob. 7.19 The result [see Fig. 7 in Ref. 28] is sketched below.

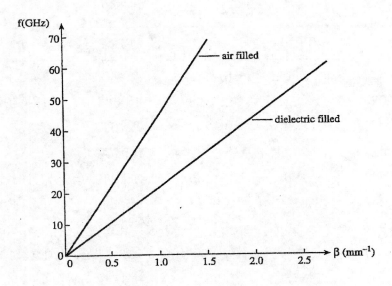

Prob. 7.20 The solution [27] is shown below.

State of cavity	TLM	Exact
Completely filled with dielectric	0.493 ns	0.490 ns
Half-filled with dielectric	0.724 ns	

CHAPTER 8

Prob. 8.1 See, for example, the subroutine in Fig. 8.2.

Prob. 8.2 (a)

$$x_1 = 1573 \times 89 + 19 (\text{mod } 10^3) = 16$$
$$x_2 = 1573 \times 16 + 19 (\text{mod } 10^3) = 187$$
$$x_3 = 1573 \times 187 + 19 (\text{mod } 10^3) = 170$$
$$x_4 = 1573 \times 170 + 19 (\text{mod } 10^3) = 429$$
$$x_5 = 1573 \times 429 + 19 (\text{mod } 10^3) = 386$$
$$x_6 = 1573 \times 836 + 19 (\text{mod } 10^3) = 47$$
$$x_7 = 1573 \times 47 + 19 (\text{mod } 10^3) = 950$$
$$x_8 = 1573 \times 950 + 19 (\text{mod } 10^3) = 369$$
$$x_9 = 1573 \times 369 + 19 (\text{mod } 10^3) = 456$$
$$x_{10} = 1573 \times 456 + 19 (\text{mod } 10^3) = 307$$

(b)

$$x_1 = 1573 \times 89 (\text{mod } 10^3) = 997$$
$$x_2 = 1573 \times 997 (\text{mod } 10^3) = 281$$
$$x_3 = 1573 \times 281 (\text{mod } 10^3) = 13$$
$$x_4 = 1573 \times 13 (\text{mod } 10^3) = 449$$
$$x_5 = 1573 \times 449 (\text{mod } 10^3) = 277$$
$$x_6 = 1573 \times 277 (\text{mod } 10^3) = 721$$
$$x_7 = 1573 \times 721 (\text{mod } 10^3) = 133$$
$$x_8 = 1573 \times 133 (\text{mod } 10^3) = 209$$
$$x_9 = 1573 \times 209 (\text{mod } 10^3) = 757$$
$$x_{10} = 1573 \times 759 (\text{mod } 10^3) = 761$$

Prob. 8.4 Using $X_{n+1} = aX_n (\text{mod } m)$, we obtain the following random numbers for cases $X_o = 1, 2, 3,$ and 4.

n	X_n	X_n	X_n	X_n
0	1	2	3	4
1	13	26	39	52
2	41	18	59	36
3	21	42	63	20
4	17	34	51	4
5	29	58	23	$\vdots$
6	57	50	43	
7	37	10	47	
8	33	2	35	
9	45	$\vdots$	7	
10	9		27	
11	53		31	
12	49		19	
13	61		55	
14	25		11	
15	5		15	
16	1		3	

Thus, the periods are 16, 8, 16, and 4 respectively.

Prob. 8.6 Let

$$f_X(x) = \sqrt{\frac{2}{\pi}}e^{-x^2/2} = \frac{2\sqrt{2}}{\sqrt{\pi}}\frac{e^{-kx}}{(1+e^{-kx})^2}$$

so that

$$F_X(x) = \frac{1}{1-e^{-kx}} - 1$$

The inverse transform is

$$X = \frac{1}{K}\ln\frac{1+U}{1-U}$$

Attaching a random sign α to this variate gives the desired variate $Z = \alpha X$. Thus we:

(1) Generate U_1 and U_2 from $(0,1)$

(2) $X \quad \leftarrow \quad \sqrt{\pi/8}\ln[(1+U_1)/(1-U_1)]$

(3) If $U_2 \leq 0.5$, deliver $Z = -X$

(4) If $U_2 > 0.5$, deliver $Z = X$

Prob. 8.7 Here $M = 5$, $a = 0$, $b = 1$. We generate the random variate as follows:

(1) Generate two uniform random variates U_1 and U_2 from $(0,1)$.

(2) Check if $U_1 \leq f_X(U_2)/M = U_2^2$.

(3) If the inequality holds, accept U_2 as the variate generated from $f_X(x)$.

(4) If the inequality is violated, reject U_1 and U_2 and repeat steps 1 to 3.

Prob. 8.8 We generate X as follows:

(1) Generate two uniform variates U_1 and U_2.

(2) Compute $Y = -2a^2(U_1 - 0.5)^2$.

(3) If $\ln U_2 \leq Y$, then accept $X = a(2U_1 - 1)$ as a normal variate.

(4) If $\ln U_2 > Y$, reject U_1 and U_2 and repeat the above process.

Prob. 8.9(a) Exact: 3.14159; the Monte Carlo results are shown below.

N	I
20	3.2544
50	3.1599
500	3.1566
10,000	3.1346

(b) Exact: 0.45970; the Monte Carlo results are shown below.

N	I
20	0.4048
50	0.4398
500	0.4512
10,000	0.4606

(c) Exact: 1.71828; the Monte Carlo results are shown below.

N	I
20	1.6259
50	1.6883
500	1.7036
10,000	1.7207

(d) Exact: 2.0; the Monte Carlo results are shown below.

N	I
20	2.0667
50	1.8986
500	1.8921
10,000	1.9566

Prob. 8.10 (a) 0.0693976, (b) 0.0

Prob. 8.11 Compare your result with the exact solution:

$$I(\pi, \pi) = \frac{\sin \pi/2 \sin \pi/2}{(\pi/2)^2} = 0.4053$$

Prob. 8.12 The finite difference equivalent is

$$\frac{W(x + \Delta, y) - W(x, y) + W(x - \Delta, y)}{\Delta^2} + \frac{W(x, y + \Delta) - W(x, y) + W(x, y - \Delta)}{\Delta^2}$$
$$+ \frac{K}{y} \frac{W(x, y + \Delta) - W(x, y - \Delta)}{2\Delta} = 0$$

assuming $\Delta x = \Delta y = \Delta$. This yields

$$W(x, y) = p_{x+} W(x + \Delta, y) + p_{x-} W(x - \Delta, y) + p_{y+} W(x, y + \Delta) + p_{y-} W(x, y - \Delta)$$

where

$$p_{x+} = p_{x-} = \frac{1}{4}, \qquad p_{y+} = \frac{1}{4} + \frac{K\Delta}{8y}, \qquad p_{y+} = \frac{1}{4} - \frac{K\Delta}{8y}$$

Prob. 8.13 Exact solution: $y = 10x$, $y(0.25) = 2.5$

For Monte Carlo solution:

$$y'' = \frac{y(x+\Delta) - 2y(x) + y(x-\Delta)}{\Delta^2} = 0$$

$$y(x) = p_{x+}y(x+\Delta) + p_{x-}y(x-\Delta)$$

where $p_{x+} = p_{x-} = 1/2$.

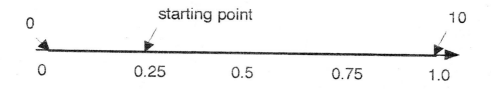

Let $0 < U < 1$ be a random number. If $U > 0.5$, $x \rightarrow x - \Delta$. If $U < 0.5$, $x \rightarrow x + \Delta$.

nth walk	V_p	m
1	0	5
2	0	1
3	10	3
4	0	1
5	0	1
6	0	3
7	10	3
8	0	1

Prob. 8.14 Compare results with exact solution:

(a) $10\frac{k}{N} = 10\frac{2}{5} = 4$ V

(b) $10\frac{k}{N} = 10\frac{7}{10} = 7$ V

(c) $10\frac{k}{N} = 10\frac{11}{20} = 5.5$ V

Prob. 8.15 Compare results with exact solution:

$$V(0.4, 0.2) = 1.1$$
$$V(0.35, 0.2) = 1.005$$
$$V(0.4, 0.15) = 1.05$$
$$V(0.45, 0.2) = 1.15$$
$$V(0.4, 0.25) = 1.15$$

Prob. 8.16 Using the Exodus method with $\Delta = 0.25$ and $N = 10^6$, we obtain

$$V(2, 2) = 42.31, \quad V(3, 3) = 16.92, \quad V(4, 4) = 4.18$$

These values may be compared with the finite difference solution with $\Delta = 0.1$ and 1000 iterations:

$$V(2, 2) = 42.59, \quad V(3, 3) = 17.72, \quad V(4, 4) = 4.33$$

Prob. 8.17 According to Example 3.4, the exact solution gives

$$V(0.5, 0.5) = 2.913$$

Using the floating random MCM, we obtain for $\Delta = 0.1$,

$$
\begin{aligned}
N = 2,500 \quad &\rightarrow \quad V(0.5, 0.5) = 3.015 \\
N = 3,000 \quad &\rightarrow \quad V(0.5, 0.5) = 3.07 \\
N = 5,000 \quad &\rightarrow \quad V(0.5, 0.5) = 2.991
\end{aligned}
$$

Prob. 8.18 From [52] of the text, we obtain the following result for $V(1.5, 1.0)$ using $\Delta = 0.05$ and 1000 iterations for the finite differene method.

Exact	Finite Difference	Fixed Walk	Exodus
23.43	22.99	23.58 ± 1.2129	23.43

Prob. 8.19 Compare your result with the exact solution: $V = 1.2$ V.

Prob. 8.20 Applying Gauss's law,

$$
\begin{aligned}
0 = \Big(&\epsilon_1 \left[\frac{V(\rho, z + \Delta) - V(\rho, z)}{\Delta} \right] \Delta + \epsilon_1 \left[\frac{V(\rho - \Delta, z) - V(\rho, z)}{\Delta} \right] \frac{\Delta}{2} \\
&+ \epsilon_2 \left[\frac{V(\rho - \Delta, z) - V(\rho, z)}{\Delta} \right] \frac{\Delta}{2} + \epsilon_2 \left[\frac{V(\rho, z - \Delta) - V(\rho, z)}{\Delta} \right] \frac{\Delta}{2} \\
&+ \epsilon_2 \left[\frac{V(\rho + \Delta, z) - V(\rho, z)}{\Delta} \right] \frac{\Delta}{2} + \epsilon_1 \left[\frac{V(\rho + \Delta, z) - V(\rho, z)}{\Delta} \right] \frac{\Delta}{2} \Big) \rho \Delta \phi
\end{aligned}
$$

Simplying this yields

$$V(\rho, z) = p_{\rho+} V(\rho + \Delta, z) + p_{\rho-} V(\rho - \Delta, z) + p_{z+} V(\rho, z + \Delta) + p_{z-} V(\rho, z - \Delta)$$

where

$$p_{\rho+} = \frac{\epsilon_1}{2(\epsilon_1 + \epsilon_2)}, \qquad p_{\rho-} = \frac{\epsilon_2}{2(\epsilon_1 + \epsilon_2)}, \qquad p_{z+} = \frac{1}{4} = p_{z-}$$

Prob. 8.21 Compare your results with the finite difference solution for $N = 500$ and $\Delta = 0.25$.

$$V(2, 10) = 65.85, \ V(5, 10) = 23.32, \ V(8, 10) = 6.4, \ V(5, 2) = 10.23,$$

$$V(5, 18) = 10.34$$

Prob. 8.22 The random (fixed) walk solution with $\Delta = 0.2$ is compared with the finite difference solution with $\Delta = 0.25$ and $N = 500$ below.

ρ	z	N	MCM	FD
5	18	1,000	10.80 ± 1.035	10.3467
		2,000	10.75 ± 0.6345	
5	10	1,000	26.14 ± 2.26	23.561
		2,000	25.98 ± 1.777	
5	2	1,000	11.16 ± 1.463	10.3467
		2,000	11.44 ± 0.8402	
10	2	1,000	2.66 ± 0.672	1.3026
		2,000	2.48 ± 0.5528	
15	2	1,000	0.58 ± 0.3866	0.1124
		2,000	0.49 ± 0.2648	

Prob. 8.23 (a) 0.33, 0.17, 0.17, 0.33

(b) 0.455, 0.045, 0.045, 0.455, 0.455

Prob. 8.24 The finite difference equivalent is

$$\Phi(x) = p_+ \Phi(x + \Delta) + p_- \Phi(x - \Delta)$$

where $p_+ = \frac{1}{2} = p_-$. Injecting 256 particles at $x = 0.25$ and dispatching particles according to p_+ and p_- leads to the result shown below.

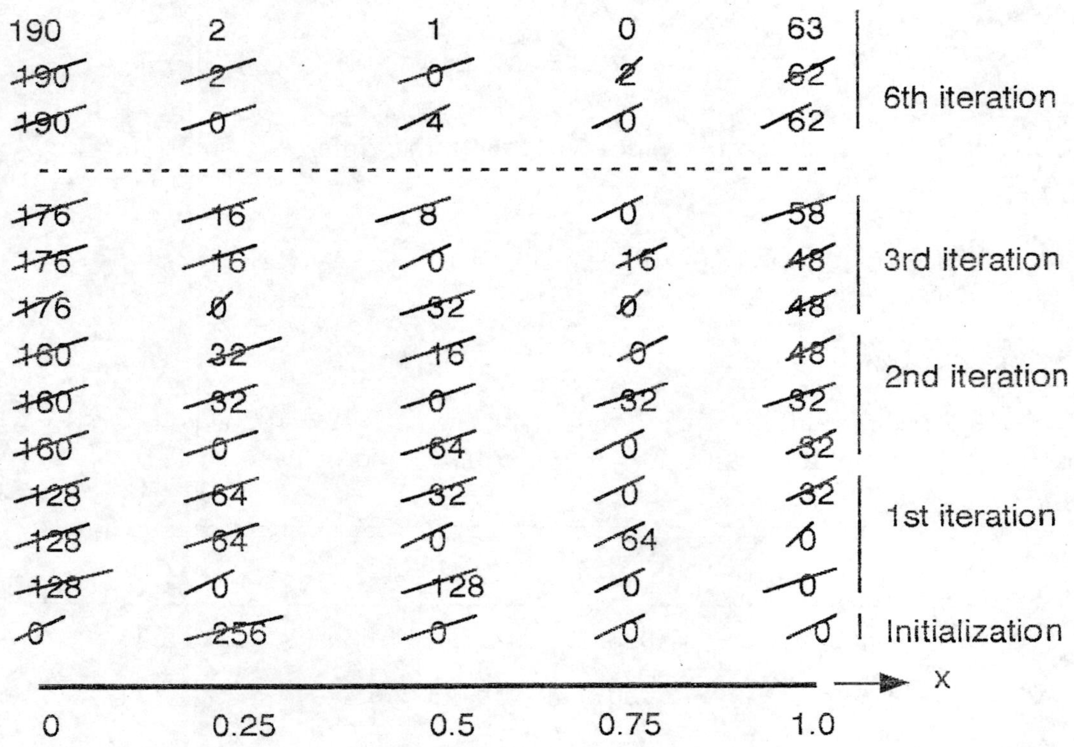

0	0.25	0.5	0.75	1.0	
190	2	1	0	63	
190	2	0	2	62	6th iteration
190	0	4	0	62	
176	16	8	0	58	
176	16	0	16	48	3rd iteration
176	0	32	0	48	
160	32	16	0	48	2nd iteration
160	32	0	32	32	
160	0	64	0	32	
128	64	32	0	32	1st iteration
128	64	0	64	0	
128	0	128	0	0	
0	256	0	0	0	Initialization

$x \longrightarrow$

The exact solution is $\Phi = 10x$. Hence, $\Phi(0.25) = 2.5$. Using the Exodus method, as shown above, we obtain after 6 iterations,

$$\Phi(0.25) = \frac{190}{126}(0) + \frac{63}{256}(10) = 2.461$$

which is only 1.56% off the exact value of 2.5.

Prob. 8.25 Dispatching the particles according to $p_{x-} = p_{x+} = p_{y-} = p_{y+} = \frac{1}{4}$ gives the result below.

After the fourth iteration, we obtain

$$N_1 = \text{no. of particles reaching 100 V side} = 10 + 21 = 31$$

Hence

$$V_4 = \sum_{k=0}^{4} p_k V_k = \frac{31}{256}(100) = 12.11$$

which is just 1.5% off the exact value of 11.928.

Prob. 8.26 Compare your result with the exact solution: $V(0.25, 0.75) = 43.20$.

Prob. 8.27 Compare your result with the finite difference solution: $V(1.0, 1.5) = 10.44$.

Prob. 8.28 Compare your result with those of fixed random walk MCM and finite element solutions:

$$V(0.2, 0.4) = 35.55 \text{ MCM}$$
$$V(0.2, 0.4) = 36.364 \text{ FEM}$$

Prob. 8.29 Node 5 is the only free node ($n_f = 1$, $n_p = 4$). The transition probability matrix is given by

$$
\mathbf{P} = \begin{array}{c} \\ 1 \\ 2 \\ 3 \\ 4 \\ 5 \end{array}
\begin{array}{c} \begin{array}{ccccc} 1 & 2 & 3 & 4 & 5 \end{array} \\
\left[\begin{array}{ccccc}
1 & 0 & 0 & 0 & 0 \\
0 & 1 & 0 & 0 & 0 \\
0 & 0 & 1 & 0 & 0 \\
0 & 0 & 0 & 1 & 0 \\
\frac{1}{4} & \frac{1}{4} & \frac{1}{4} & \frac{1}{4} & 0
\end{array} \right]
\end{array}
$$

It is evident that

$$Q = 0, \qquad N = (I - Q)^{-1} = I, \qquad R = [\, \frac{1}{4} \quad \frac{1}{4} \quad \frac{1}{4} \quad \frac{1}{4} \,]$$

$$B = NR = [\, \frac{1}{4} \quad \frac{1}{4} \quad \frac{1}{4} \quad \frac{1}{4} \,]$$

$$V_f = BV_p = [\, \frac{1}{4} \quad \frac{1}{4} \quad \frac{1}{4} \quad \frac{1}{4} \,] \begin{bmatrix} V_1 \\ V_2 \\ V_3 \\ V_4 \end{bmatrix}$$

$$V_5 = \frac{1}{4}[100 + 0 + 0 + 0] = 25.0$$

which agrees with the exact solution.

Prob. 8.30 From [56], the Markov chain solution at $(x, y) = (1.5, 1.0)$ with $\Delta = 0.1$ is $V = 23.41$.

Prob. 8.31 From [56], the Markov chain solution is shown below.

Node ρ	z	Markov chain	Exodus method	Finite difference
5	18	11.3931	11.438	11.474
5	10	27.4727	27.816	27.869
5	2	12.3099	12.179	12.128
10	2	2.4481	2.3523	2.3421
15	2	0.4684	0.38423	0.3965

CHAPTER 9

Prob. 9.1 Since

$$\frac{\partial^2 V_i}{\partial y^2} + \frac{1}{h^2}[V_{i-1}(y) - 2V_i(y) + V_{i+1}(y)] = 0$$

inserting $i = 1$ and Dirichlet condition $V_0 = 0$, we get

$$\frac{\partial^2 V_1}{\partial y^2} + \frac{1}{h^2}[V_0(y) - 2V_1(y) + V_2(y)] = 0$$

which implies that $p_\ell = -2$.

For Neumann condition, $V_0 = V_1$ which yields

$$\frac{\partial^2 V_1}{\partial y^2} + \frac{1}{h^2}[-V_1(y) + V_2(y)] = 0$$

This implies that $p_\ell = -1$.

Prob. 9.3 As in Example 9.2, λ_k remain the same.

$$\lambda_k = 2\sin\phi_k/2 \tag{1}$$

Since

$$t_i^{(k)} = A_k e^{ji\phi_k} + B_k e^{-ji\phi_k}$$

$$t_0^{(k)} - t_1^{(k)} = 0 \qquad \rightarrow \qquad A_k + B_k - A_k e^{j\phi_k} - B_k e^{-j\phi_k} = 0 \tag{2}$$

$$t_{N+1}^{(k)} = 0 \qquad \rightarrow \qquad A_k e^{j(N+1)\phi_k} + B_k e^{-j(N+1)\phi_k} = 0 \tag{3}$$

Putting these together, we obtain

$$\begin{bmatrix} 1 - e^{j\phi_k} & 1 - e^{-j\phi_k} \\ e^{j(N+1)\phi_k} & e^{-j(N+1)\phi_k} \end{bmatrix} \begin{bmatrix} A_k \\ B_k \end{bmatrix} = 0$$

For nontrivial solution,

$$e^{-j(N+1)\phi_k} - e^{-jN\phi_k} - e^{j(N+1)\phi_k} + e^{jN\phi_k} = 0$$

which leads to

$$\sin(N+1)\phi_k - \sin N\phi_k = 0$$

or

$$\phi_k = \frac{k - 1/2}{N + 1/2}\pi \tag{4}$$

From (2) and (3),

$$(1 - e^{j\phi_k})A_k = -(1 - e^{-j\phi_k})B_k \quad \rightarrow \quad e^{j\phi_k/2}A_k = e^{-j\phi_k/2}B_k$$

$$t_i^{(k)} = A_k \cos(i - 1/2)\phi_k$$

$$T_{ij} = \sqrt{\frac{2}{N + 1/2}} \cos \frac{(i - 0.5)(k - 0.5)}{N + 0.5}$$

$$\lambda_k = 2\sin\left(\frac{k - 1/2}{N + 1/2}\right)\pi$$

Prob. 9.4 By symmetry the problem can be reduced to

$$\nabla^2 \Phi = 0$$

subject to

$$\Phi(0, y) = \Phi(1, y) = 0, \quad \Phi_x(0, y) = 0, \quad \Phi(x, b) = \sin \pi x, \quad \Phi(x, y) = \Phi(x, -y)$$

Prob. 9.5 Follow Example 9.1 except that $T = T_{NN}$.

Prob. 9.7 The MOL solution is compared with the exact solution below.

(ρ, z)	Exact Solution	MOL Solution
(0, 0.25)	5.375	5.370
(0, 0.5)	7.213	7.196
(0, 0.75)	5.375	5.370
(0.125, 0.5)	7.433	7.415
(0.25, 0.5)	8.065	8.046

Prob. 9.8 Due to symmetry, We consider half of solution region. The result [11] is shown below.

157

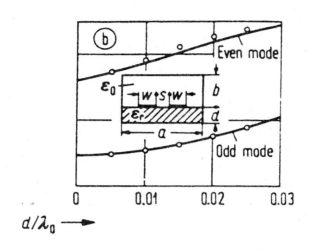